Research in action

CERES

RESEARCH SCHOOL FOR RESOURCE STUDIES FOR DEVELOPMENT

This publication was made possible in part with support of BO-09. BO-09 is a research programme supported by the Dutch Ministry of Agriculture, Nature and Food Quality focusing on the creation a 'dynamic green knowledge system' in the Netherlands by investigating ways to creating synergy between research, education and business & industry.

Research in action

Theories and practices for innovation and social change

edited by:
Conny Almekinders
Leni Beukema
Coyan Tromp

Mansholt publication series - Volume 6

Wageningen Academic
Publishers

Cover figure based on
The learning cycle by J. Hagmann, K. Murwira, P. Ficarelli, E. Chuma, J. Ramaru and H. Peter, 2003. PICOTEAM – institute for people, innovation and change in organisations. www.picoteam.org.

ISBN 978-90-8686-087-6

ISSN 1871-9309

First published, 2009

Wageningen Academic Publishers
The Netherlands, 2009

Mansholt Publication Series

The Mansholt Publication Series (MPS) contains peer-reviewed textbooks, conference proceedings and thematic publications focussing on social changes and control processes in rural areas and (agri)food chains as well as the institutional contexts in which these changes and processes take place. MPS provides a platform for researchers and educators who would like to increase the quality, status and international exposure of their teaching materials or of their research output.

The Series is named after Sicco Mansholt (1908-1995), who was Minister of Agriculture in The Netherlands from 1945 until 1958. From 1958 until 1972 he was Commissioner of Agriculture and Vice-President of the European Commission.

MPS is supported by the Mansholt Graduate School of Social Sciences (MG3S) and CERES Research School for Resource Studies for Development. The quality and contents of the Series is monitored by an interdisciplinary editorial board. MPS is published and marketed internationally by Wageningen Academic Publishers.

Contents

Part 2: Theoretical reflections

Preface

The advantage of a national research school is that people from different disciplines and corners of the research field can meet relatively easily and get to know about each others work. That is how this publication has come about.

A number of years ago, the graduate school CERES brought together a number of researchers each of whom had an interest in '*Research in action*': research focused on change, research in which the relationship with the researched is one of collaboration, research that is interdisciplinary in character. The researchers were partly from Wageningen University (Gemma van der Haar, Gerard Verschoor, Conny Almekinders) and partly from the University of Utrecht (Ben Valkenbrug, Coyan Tromp, Leni Beukema). In addition, we had different disciplinary backgrounds: interdisciplinary social sciences, development studies, philosophy, and technology studies. With these diverse backgrounds we started discussing the possibility of publishing a book about *research in action* which would principally serve PhD students. That idea had very fruitful results: everybody had his or her own orientation, issues and questions of which the others learned. However, it also meant we had a long way to go: it took us time to find our commonalities and differences, value these and arrive at a joint idea about what *research in action* could stand for and what it had to offer researchers involved in a PhD project. Also, the discussions and joint collaboration on this innovative project joint project meant in practice that we had to combine this with our day-to-day worries at work and in our life. Not everybody was able to spend the same amount of time on the project.

It is therefore with some degree of pride that we now can present this publication. We have not only wanted to share the practices of *research in action* in different contexts, but also wanted to link these practices to theoretical and methodological choices which each researcher has to make in the design and implementation of research. Instead of taking a prescriptive approach, we have – in line with *research in action* itself – chosen an open approach: what can you think of when developing research, what are the opportunities, which dilemmas might you face, etc. In this way we want to contribute to the current discussion about *research in action*. We also take an open approach here: the book is a reflection of our own development as researchers. Constructive comments about what others consider more or less successful in this sense will be very much appreciated. For us the book is a new start to the discussion we had in our own network and which we enjoyed very much.

As with everything, this book also comes with words of thanks. First of all to Lolita van Toledo at CERES who brought us together, strongly supported by the Daily Board of CERES Ad van Eldijk and Ton Dietz. Also the current CERES board, especially in the person of Han van Dijk, has always believed in a valuable end product of our discussions and has contributed financially to this publication. Arjen Wals, in his role as former leader of one of the LNV Policy Research Clusters (BO-09) and as one of the Series Editors of the Mansholt Publication Series, has been very helpful in exploring and pursuing the options to publish this book. Last but not least a word of thanks to our 'masters', Paul Richards, Wageningen and Harry Coenen (in memoriam), Utrecht, whose input and influence on our thinking should not be underestimated.

Wageningen, Driebergen, Amsterdam, August 2008

Conny Almekinders, Leni Beukema, Coyan Tromp

1. Research in action

Options, choices and challenges

Ben Valkenburg, Leni Beukema, Conny Almekinders and Coyan Tromp

1.1 A first introduction to research in action

Many undergraduates and PhD students are interested in research that does more than contributing to scientific knowledge. They see having an impact and contributing to change and emancipation as important goals of their research and are therefore looking for possibilities to do their research in ways that directly contribute to these goals: they want to do research in action.

To do so, however, they have to answer a lot of questions. How do they deal with the questions and goals of the people or organisations that supply the money for research? Who will be the primary stakeholders and what should be their role in the research project? What methods of data gathering can be used? How can the results of research be given back to the people being studied? How does one deal with power relations and opposing interests in the research? These and other questions often have to be answered in a situation where many things have already been decided before the research starts. In many cases the period and time for research is determined and, often, too short for all the things the researcher would like to do. Sometimes, especially for PhD students, the research plan has been made beforehand. And, last but not least, sometimes the situation under study just does not offer the required conditions for research in action to be successful.

The aim of this book is to assist undergraduates and PhD students who want to do research in action. We do not offer recipes, telling the reader which ingredients he or she should use and how to make a dish out of them. Neither will we offer a travel schedule, telling the reader how to get from a to b. Our starting point is that undergraduates and PhD students must be able to work with the book in very different situations and under different conditions. Our aim is to present ideas, experiences, reflections and concepts on research in action that can help them to formulate their own answers to the many questions they are facing.

We have chosen '*Research in action*' as the title of this book, to cover a whole range of research strategies that aim to bring about innovation and social change. Among the types of research that can be included under research in action are: participatory action research, participatory extension approach, participatory learning and action research, exemplary action research, collaborative research, co-operative inquiry and various forms of intervention research. Many of these types of research can be found in different handbooks on action research (see e.g. Reason and Bradbury, 2001; Greenwood and Levin, 2007). However, the desire to bring about social change and innovation is not the exclusive domain of action researchers.

More and more we see researchers struggling with the question of implementation of scientific knowledge in the field of study and its consequences for cooperation with different stakeholders, as for instance in agricultural science or development studies. In this way research and action are becoming increasingly linked.

Within the mainstream of research in action we see a rather principal divide. Some researchers regard research in action mainly as just one more type of research. In that case, it is viewed as a research strategy or methodology, in addition to for instance an experiment, a survey and a literature study. Research in action would then be viewed more or less as an equivalent of field research. For others, research in action is not merely another research strategy, but implies a different outlook on what science is or should be. As such, it not only entails particular research methods and techniques, but also more fundamental principles about how scientific research can best be organised and framed. This more fundamental basis of scientific and philosophical principles is called a *paradigm.*[1] Because we want to provide room for both views, i.e. research in action viewed as research strategy and as paradigm, we have opted for '*Research in action*', since it can function as an umbrella covering the various possible viewpoints, while emphasising the prominent role of the action with co-researchers in the research.

The issue of the co-researchers unveils the need for a second clarification of the terminology used in this book, concerning the terms used to indicate the researched party. Within conventional research, in which knowledge production is practiced in the scientific domain and is viewed as a prerogative of scientists only, the researched are treated mainly as passive sources of information. Within research in action, the intention is to change this role, and involve the researched as active participators in the research. In these action-oriented approaches, knowledge is developed in continuous dialogue between scientists and non-scientists. The underlying view of knowledge development is that the researched party, or co-researchers, can contribute valuable contextual knowledge of the actions and systems under study that is supplementary and even indispensable to the theoretical and methodological knowledge of the scientists. This means that the researched are given a serious role within the process of data collection, interpretation and analysis, to be able to continuously connect practice to theory and vice versa so as to be able to bring about the necessary interaction between the two.

In practice however, the ideal of involving the researched as co-researchers is difficult to attain on all fronts. Not everybody can participate all the time and with the same intensity in all the phases of the research process. The aim is to get as close as possible to the ideal, but even then we should note that the question of what 'the ideal' is will be a matter of weighing pro's and con's of active participation – as we will come to see in the following

[1] Complex, theoretically laden terms like this one are taken up in a glossary in the back, to enable the reader to review the definitions when stumbling upon them in the text.

chapters. In the actual research practice there will be inevitable agreement on compromises at variance with the 'ideal'. Since it would therefore be rather presumptuous to talk about co-researchers everywhere, the researched party will sometimes be referred to as 'the researched' and sometimes as co-researchers, depending on what seems to be legitimised within the specific context.

In addition to considering co-researchers as essential partners in the knowledge generation, this view on knowledge development implies a different relationship between those involved in the research. In traditional science, the aim is to generate value-free results which means that the researcher has to try and keep the researched party at a distance. In research in action the cooperation between researchers and co-researchers is taken as a starting point, a relationship of trust with those involved in the research is a *sine qua non* for research; without trust the results of the (interaction in the) research project will not be trustworthy (see also Beukema, Chapter 11).

In the remainder of the chapter we will introduce some more core elements of what we call research in action (paragraph 1.2). We then introduce the main perspectives from which we will develop our ideas, experiences, reflections and concepts. We start with the notion of plurality, where we recognise that different perspectives and approaches exist within science. Within these different approaches we indicate how we see the place of research in action (paragraph 1.3). Knowledge of this methodological discussion on science and scientific knowledge is helpful to explicate assumptions and develop the position of the researcher as part of the research activities. The second perspective is diversity within the broad range of research in action itself and the impact of the contextuality on this kind of research (paragraph 1.4). There is not one 'form' of research in action; there are many. This also follows logically from the consideration that the context is important and thus should play an important role in the decisions the researcher makes about the organisation of the research project.

The third perspective is the researcher as a person (paragraph 1.5). The decisions on a research project have consequences not only for the way research is done, but for the researcher himself and for other actors concerned. These consequences touch on the relationship with people that commission the research project and deliver the money, on the demands on the researcher him- or herself, and on his or her relationships with colleagues in the academic world.

In a later chapter (Tromp *et al.*, Chapter 12) we will come back to these core elements and perspectives. We will review them on the basis of the diverse range of experiences being presented in the following chapters of this book. At the end of this chapter we introduce these experiences (paragraph 1.6).

1.2 Core elements

As mentioned above, research in action can be diverse in form and can be found in a range of practices. Nevertheless, at the core of this diversity there are five common elements. Together these elements can serve as a preliminary 'open' description of research in action. We present them as a first point of reference for the reader.

First, the effort to achieve social change and emancipation means that the researcher wants the scientific knowledge and the process in which it is developed to be an important resource for the people under study to improve their situation. He or she not only tries to find out what reality is, but also actively engages in experimentation to find out how the reality can be changed. The researcher tries to answer the question on change from the perspective of the people at the 'short end of things'. These are primary stakeholders, i.e. the people that should be helped to use scientific knowledge to strengthen their resources and to use these resources to change their situation.

This brings in a second element: the researcher not only wants to know what the people under study say, what their characteristics are and how these two are related. He or she also wants to understand the actions of people, their underlying interpretations, knowledge and strategies. Moreover, the researcher tries to get a grip on the (contextual) conditions for and consequences of these actions. Last but not least, the researcher wants to develop possibilities, preferably together with the people involved, to direct these actions towards change. Take, for example, the effort to try and understand how people are dealing with HIV and AIDS infection. A researcher in action would want to know: for people with HIV and AIDS how does it affect the way they see their position in the community, what is their position in the community, and what are the conditions for actively participating in the community? The aim would be to find possibilities to improve their situation. In this second element, the confluence of action and research comes together: the design and implementation of action generates knowledge that contributes to the overall understanding of the situation of the people involved. This knowledge is of direct use for these people (i.e. the co-researchers), but also – usually documented in a different form – contributes to the body of scientific knowledge.

Often a third element is added to the two former ones: the aim to achieve a reciprocal relationship between the researcher and the people under study, i.e. the researched. In more traditional concepts of science the relationship between researcher and the researched is one-sided. A sharp distinction is made between scientific knowledge and common-sense knowledge. The role of the people under study is limited to providing data. The scientist is the one who uses these data for a scientific analysis. The communication between the researcher and the people under study is limited to the process of data gathering and to the moment that the researcher delivers the results (in the form of a research report, including his or her recommendations). In opposition to this traditional concept, many researchers in action argue that a good analysis of what people do is only possible if these people themselves can and want to contribute to it. 'Can' implies that the researcher sees and treats these people as co-researchers. 'Want to' implies that they must sufficiently trust the researcher and the project he or she is undertaking. Only then are they willing to 'really' tell the researcher what they are doing and why they do it. 'Can' and 'want to' together require a dialogical relationship between the researcher and the people under study, instead of it being an extractive, one-sided relationship.

A fourth element is related to the above. The situation under study is approached in all its complexity: a mixture of individual action, collective action, and social developments and conditions. Research in action starts from an inter- or transdisciplinary perspective. The traditional distinctions - not only between the different social scientific disciplines, but also between the social and natural scientific disciplines - are not taken for granted. For instance, when agriculturists are presenting soil-laboratory results to small-scale farmers they not only need their technical skills, they also need communication skills to combine their expert knowledge with the farmers' contextual knowledge of their fields. Moreover, they need social scientific knowledge such as insight into relationships between different farmers, cultural differences between farmers and between themselves and the farmers to interpret the interaction that is taking place (see Ramaru *et al.*, Chapter 3). Depending on the issues under study, researchers in action try to use knowledge from different disciplines (interdisciplinary knowledge) and combine it with other, non-scientific knowledge like common sense knowledge in order to generate transdisciplinary knowledge and so understand the reality under study in all its complexity.

A fifth core element for science dealing with people in action is the importance of reflexivity, i.e. the object of science itself has learned to think along scientific lines. The expert only has a relative advantage over the well-informed lay person. This is especially true in today's situation, in which the social space for action (that is: the social world around us, which serves as a basis for reflection on our individual identity and actions) can for many of us be expanded to the whole world with one click of the mouse on our personal computers. In this process, the meaning of knowledge has fundamentally changed. Knowledge can no longer be regarded as a cumulative source of certainties. Knowledge is being permanently used, discussed and developed in everyday practice, i.e. it involves a reciprocal process, involving many actors, taking place in the context of everyday interaction. If the contexts are different, this reciprocal process may be different too. General statements can only be expressed in modest terms, and certainly not in terms of de-contextualised laws of social behaviour and predictions.

This changed meaning of knowledge is at the same time an important factor in, and the outcome of, processes of individualisation. Based on a permanent flow of new knowledge, and in different contexts, actors develop their individual identity and actions. Individual identity has become a reflexive project. The reflexive monitoring of action means that actors use their knowledge of themselves and their social context to translate their motivations and intentions into actions, and at the same time to monitor this process. In this sense many people in today's world are more competent actors than ever, that is, able to steer and give direction to their own lives (Valkenburg and Lind, 2001).

1.3 Methodological plurality

In the methodological discussion different answers are given to the question 'what is scientific knowledge and how can it be developed?' This discussion and the different answers given are as old as science itself: is the goal of social science to study social reality, *and* change it? Or should the goal of changing social reality be rejected in science? In the old days this discussion was a matter of 'taking a political position in the debate'. The methodological arguments underlying these positions have been elaborated more recently.

Textbooks for students in social sciences often distinguish three research approaches: empirical-analytical research, interpretative research and critical-emancipatory research. (see e.g. 't Hart *et al.*, 1996: 75 ff) Scientists of the empirical-analytical approach (still the dominant view in social science in the Netherlands at the moment) make a plea for a methodology that comes as close to the successful natural sciences as possible. The general opinion of these scientists is that objectivity, reproducibility and controllability and control of the research situation can also be realised in the social sciences. Scientists of the second approach, the interpretative approach (found in much of the ethnographical research), see the main point of social sciences as the understanding of the motives, beliefs and intentions of people in their behaviour and living together. Some interpretative social scientists follow formal methodological rules, others speak of substantive methodology, where the content of the observed, its interpretation and the method of participatory observation are described in one narrative story. Scientists of the third approach, critical-emancipatory research, aim not only for knowledge development but also for direct personal and societal change through their research. They see research not as a value-free activity of social scientists, but as a project of scientists and lay persons in an attempt to create a world with less inequality and for an expansion of people's (access to) resources that can help them to give direction to their own lives.

In the context of this book on research in action, we believe that an important watershed in this debate is the extent to which social science should or could conform to the methodological principals of the natural sciences (a traditional empirical-analytical approach or not). In the empirical-analytical tradition the researcher formulates his or her theoretical ideas about how the world functions and looks at the empirical reality to see whether or not these ideas are correct. The empirical observations are decisive for the theoretical ideas. By permanently testing these theoretical ideas, they are developed into general laws that presumably 'shape' the natural reality: gravity makes an apple fall. The object under study, the apple, has no influence on what happens. It has no ideas and interpretations of the situation, and makes no decisions about it. The same goes for the context: the apple will fall in Holland, as it will fall in South Africa. The laws of gravity are the same everywhere. Most empirical-analytical scientists are convinced that this scientific model can and should be used as an ideal model for the social sciences too. Others disagree, since people and apples are different things. These researchers therefore prefer a model in which there is (also) room for an interpretative and critical emancipatory approach. The underlying rationale is that people think about their own reality, and have interpretations that shape their actions; in other words they are reflexive persons. With their reflexive actions they change their reality, albeit not in total freedom and under self-chosen conditions. This means that in social reality there are no general (let alone universal) laws. Furthermore, the social researcher is not an outside spectator. The researcher himself/herself is an actor in the reality under study. The research activities have an impact on this reality. When a researcher reports, for example, that eighty percent of the Dutch (or French, Germans, etc.) have negative ideas about foreigners, this will generate publicity. This may very well stimulate some Dutchmen (or Frenchmen, etc.), to critically assess their own ideas and to actively change these ideas. This process can lead to a situation in which, after some time, only seventy percent have negative ideas. In the terminology used by the well-known English social scientist Anthony Giddens, this is described as follows: social scientific knowledge spirals in and out of social reality, and in this process knowledge and reality influence each other.

These different approaches can also be found within research in action. Some researchers in action see research in action primarily as a means to give the results of their research 'back' to social actors. In this view, the methodological and theoretical choices of the research are only a concern of the scientists themselves. They see research in action as one of the many ways of data gathering that can be chosen. These researchers can be found in a small part of the empirical-analytical and mainly in the interpretative stream. Other researchers in action (from the interpretative, but mostly from the critical-emancipatory stream) more often conform to the idea that combining research with action that aims at change has far-reaching consequences for the meaning of objectivity, validity, adequacy and generalisation. They see research in action as an alternative paradigm: a concept of social science and generation of scientific knowledge that is fundamentally different from other, more analytically-based concepts. Between these two poles a broad range of concepts of research in action can be found.

1.4 Diversity and contextuality

As stated above, research in action takes a stand in the general methodological debate in social science. It has also become clear that not all researchers in action take the same approach. The five elements we introduced in paragraph 1.2 can be translated into the practice of research in many different ways. There are, in other words, many different forms of research in action. This goes, for example, for the active role of the people under study. The modest way of actively involving the researched is to regularly inform them about the activities of the researcher. Their reactions can be used as 'data' by the researcher during the research project. Sometimes their role is more extensive and involving. The researcher supplies the people under study with the instruments to investigate their own reality and to deliver the results as input for the research project. In such situations the people under study become co-researchers. Other researchers try to cooperate with the people under study in terms of equivalence. The people under study are not only seen as co-researchers, they are also asked to assess the knowledge the researcher produces. The analyses are seen as adequate if both the researcher and the people under study agree on it. This principle is called reciprocal adequacy.

A comparable range of attitudes can be distinguished with respect to the relationship of trust that is needed in action-oriented research. Sometimes this relationship is primarily seen as a personal relationship between the researcher and the people under study. The researcher tries to develop an open, honest and constructive relationship. At the other end of this continuum the argument is that trust can only be developed if the people under study have a say in all aspects of the research project from the beginning of the study onward (formulating issues, questions and goals, methods to be used, analyses of data and results). On the continuum between these poles, many other variants can be found.

Variety also goes for the way research in action contributes to change. Some see change as the principal criterion: if the situation under study has changed, research has been successful. Development of knowledge is considered instrumental to this process of change. Others see the development of knowledge as the principal purpose and criterion of success. Change is seen as a potential result, after the research project has been finished. In this case the researcher gives his or her recommendations at the end; what people do with these recommendations is not part of the research process. Sometimes change and the development of knowledge are seen to be connected in a reciprocal way: knowledge is needed to change a situation and changing this situation together with the people under study is seen as an important way of generating further knowledge.

The varieties above are not only based on the ideas and concepts of the researcher. They are also, to a certain extent, a result of the situation in which research is done. Almost by definition research is a 'public affair'. When employees or communities are, for example, actively involved in a research project to stimulate democratic structures in their company or village, management and leaders will soon know. Via the researcher, managers and leaders can inform themselves, not only on the role of employees or community members as co-researchers, but also of what they are saying in this role. In an open situation, where all actors are convinced that democratic structures should be strengthened, this would be no problem. These really open situations are exceptional, though. In many cases the involvement of employees of a company or institution in such a research can lead to sanctions. In situations, for example in countries where interests are strongly opposed and protection of employees is limited by law, or non-existent, these sanctions can be severe. The researcher must take this into account, also in the way he cooperates with the people under study. In other words, research in action purposefully aims to line up with the context and this makes the practice of research in action diverse, not only in terms of the research format, but also in terms of the roles of the researcher and the people under study, and of contents of the research project. If the researcher is studying democracy in an open situation, the content will be different, as compared to a situation with strongly opposed interests. In the first situation the researcher might study how the actors concerned can cooperate. In the second situation it could very well be that the choice would be made to limit the research to the perspective of one of the actors concerned. In the end, the knowledge generated in these two situations is also likely to be different.

1.5 The researcher as a human being

The way research in action is carried out is not only connected to the decisions on the research format and to methodological arguments. Research is done by researchers, i.e. human beings, in a specific context and under specific conditions, and this too has its influence.

The researcher is not only guided by rational arguments, but also by personal considerations. Some aspects of research in action may be convenient, others may not. An intensive cooperation with the people under study, in which these people have a lot to say on the research process, is only possible if the researcher can and wants to deal with their input and involvement. The researcher must constantly keep his or her ears open for what people think and say, must be willing to give his or her own ideas and opinions as input for debate, and must be able to live with the uncertainties of an open-ended research project that takes shape 'under construction'. Some researchers look forward to this sort of challenge; others get the shakes at the mere thought of it. In other words: the way research in action is done, must - besides other considerations - be convenient for the researcher as a person. This also means that decisions during the research process should be considered as regards their consequences for the researcher, before they are made.

Besides the researcher there are other human beings involved as well: the people under study, the people that commission the project, colleagues in the academic world, friends and family. All these people bring in their own personal peculiarities and considerations; many of them must be dealt with. This creates a network of people that to a certain extent also shapes the conditions that are relevant for the decisions that are to be made in relation to the research. For the research project, some influences might be desirable, while others may not. The clearest example is that of the people who commission and finance the research project. If research is only possible for a limited period of time, it makes little sense to try to involve all actors concerned in the project over a longer period. Of course, in this situation the researcher can take a principal stand. If this means, however, that he or she cannot do any research at all, both the researcher and the people that might benefit from his research end up empty- handed. In this situation, for good reasons, the decisions are often taken by reasoning the other way around: how can we get the most out of the research and set the conditions optimally to our advantage?

1.6 The content of the book

After this introduction we start with the practice of research and the possibilities and difficulties researchers face when trying to perform research in action. As the aim of the book is to show the broad spectrum of research in action, the cases presented are diverse and situated in a wide variety of contexts. Through these cases we try to clarify the decisions made by researchers, as well as the background to these decisions. The cases are on two levels. Firstly, there are research reports: stories of research that can be told from start to finish. Reconstruction of these researches gives insight into the research process as a whole, different stages within this process, different roles of researchers, co-researchers and others involved and into the way in which results of the research were formulated and used (within and after the research). Secondly, thematic cases highlight certain aspects of the research under study that can be of particular interest when setting up a research. These cases are particularly relevant for those students who are facing a comparable situation.

The first practice we encounter is on the intersection of environment and health in Flanders, Belgium. Hans Keune, Dominique Aerts, Ludwine Casteleyn, Dirk Wildemeersch and Lieve Goorden describe how within the Flemish Center for Environment and Health, natural and social scientists team up with policymakers to enhance science-based policy development. They sketch a picture of the process in which the scientists investigate the relation between environmental and human health, and interpret the biomonitoring data on environmental pollutants and health effects. These findings are shared in the discussion with policy makers and further investigated and discussed in an action plan, which also contains a decision model to formulate priorities. After deliberation the parties further develop the outcomes into future policy strategies on environment and human health.

Joe Ramaru, Jürgen Hagmann, Zacharia Mamabolo, and Michael Netshivhodza describe the experiences of a programme that used a Participatory Extension Approach in the Limpopo Province in South Africa. The objective of the programme was to improve service supply and bring about innovation in agricultural practices of small-scale farmers. For this purpose extension officers who had been trained in facilitation skills, and soil scientists engaged with farmers in what became a mutual learning process and a blending of different types of knowledge. The researchers learned from the farmers about the way they looked at the soil and managed its fertility. This gave them the entry point to discuss options for improvement that they jointly explored through experimentation. The chapter gives a narrative description of what transpired during the interaction between researchers, extension officers and farmers, and how this significantly contributed to the farmers' capacities to deal with their own problems.

Leni Beukema and Ben Valkenburg present a study on the transformations care organisations are going through in their effort to implement a demand-driven care system for the elderly in the Netherlands. They describe the process in which a 'regular' research project became an action research project when the actual realisation of change became an issue. Secondly, they describe the steps they have taken in the different phases of the research. They show that, although the cycle of action research seems logical and simple, in reality steps constantly intermingle. They focus particularly on the relationship between researchers and different co-researchers and how they dealt with the different perspectives of different co-researchers involved in the process. A good relationship is important, Beukema and Valkenburg argue, both with regard to the implementation of change, since change requires mutual trust and to finding adequate knowledge on demand-driven care, for this trust is also necessary for people to talk and act as freely as possible. The authors finish by investigating the questions and demands such a relationship poses for organisations and their members.

Conny Almekinders, Jet Proost and Jürgen Hagmann describe the experiences of an externally financed programme for foreign PhD candidates in Wageningen University. The programme became a learning experiment on the introduction of competence development for more development and change-oriented research in developing countries that emphasised soft skill development, including those that aim to enhance reflexivity. The contribution offers an outline of the structure of the programme and the tensions that arose around different interpretations of the role of academic training. Development of trust and mutual understanding were essential in aligning the course of the programme. The reflexive mode in which the experiences are presented brings out learning that is relevant for curriculum development and institutional change related to forming a new kind of professional.

Rudi Roose and Maria De Bie reflect on the emancipatory character of action research by analysing their research on youth care in Waasland, Belgium. It is often claimed that emancipatory research should be viewed as democratic research: rather than the implementation of changes desired by the researcher in social reality, it is about societal change induced by a democratic research process in which not only the researcher but also policymakers and clients take an active part. After a short description of the background of their research they analyse the conditions for such an emancipatory research and the possibilities *and* difficulties of realising such conditions in practice, such as dealing with the question of involvement and distance during the research, with power relations in the network between professional, client and researcher, acknowledging the fact that social change takes time. They conclude that in the end, it is the task of the action researcher himself to actively participate in realising these conditions.

Gemma van der Haar examines the preconditions and limitations to collaborative research in studying political conflict. In her own case, this involved a study in the Mexican province of Chiapas on competing claims to governance by the national state on the one hand, and the Zapatistas on the other. Van der Haar acknowledges that collaboration within research deserves serious attention, for one reason because it forms a precondition to gain access to the research population. But also from a moral or ethical and a political point of view, it can be important to try and set up a dialogue with the parties involved in the study. However, it might be hard if not impossible particularly in the study of conflict, to induce a research situation in which members of the opposing parties actively want, or are allowed, to participate. From the perspective of the researched, commitment to the group's cause is one of the main criteria for assessing the legitimacy and trustworthiness of the researcher. Van der Haar points out that a researcher might want to refrain from an exclusive engagement to protect his or her independent position and to do justice to the need to explore multiple perspectives on the situation.

In Gerard Verschoor's contribution we are also confronted with multiple perspectives. This time, the focus is on the various, sometimes incommensurable evaluative frameworks that stakeholders in research use to assess what benefits the common good. Taking as an example the debate about the implementation of the technology concerning genetically manipulated maize in Mexico, he describes the arguments used by proponents and opponents, and shows how these are based on different 'regimes of justification'. The five most commonly used regimes to back up the argumentations subsequently refer to means – ends relationships (1) from an economical point of view, (2) from the perspective of efficiency or (3) collective welfare, either (4) emphasising the importance of local history, culture and tradition or (5) stressing the environmental effects of technologies. Innovation processes like those concerning genetically manipulated organisms (GMOs) are characterised by high levels of complexity and uncertainty. Therefore, Verschoor thinks that the social movement against genetically modified maize should stop doing what it doesn't manage to do (taking the high moral ground by pointing to the certainties of its evaluative frameworks) and instead emphasise what it does extremely well: pointing at the uncertainties (ecological, social, economic) involved in the introduction of such radical new technologies.

In part II of the book, we examine the philosophy and theory underlying a research paradigm or research strategies oriented towards the valorisation and implementation of scientific knowledge. To be able to transfer the knowledge (back) into practice, to make it of value for specific communities or of use for our global society, the research has to be put into action. Indeed, it must be research in action!

To begin with, and with a view to retrieving arguments in favour of a paradigm shift, Gerard Verschoor presents a birds-eye perspective of the history of the social sciences. He indicates where the important turns were taken, and which have resulted in today's situation: a social science in which the position of the researcher is preferably one that is 'detached' from its subjects in order to guarantee objectivity, and in which the role of the subject is marginal. He takes us through the Enlightenment – via the Greeks - when the idea that science could explain, improve and control the world took root, to the Romanticists and the reactions to the various modern forms of (scientistic, empiricist, positivist) social science by microsociology, the critique of ideology, feminism, and post-structuralisms. These reactions, Verschoor argues, make visible what the Enlightenment project tried (with little success) to repress. More importantly, he pleads, what has been made visible as well is the possibility of fruitful engagement between scientists and those outside the scientific arena who have thus far not been called upon and not been given a voice in the 'crafting of reality'.

Next, Coyan Tromp reflects upon the presuppositions underlying the scientific enterprise in general. To get a clear view on those assumptions, she investigates the paradox our knowledge society seems to be facing: while the potential for innovation and social change is there, the transmission of knowledge to society is far from optimal. The reason appears to be the rather narrow concept of rationality that is usually guiding scientific research, which is only limited in focus and does not take into account the action dimension that is needed to bridge the gap between science and society. Tromp argues that we should broaden our scope and not exclusively focus on our cognitive-instrumental interests in knowledge, but also on our more practical, moral and existential needs. Only then, she claims, can science offer real perspectives on innovation and social change, i.e. the vision, action strategies and possible solutions that are needed to deal with the complex questions our society is facing.

In her contribution, Leni Beukema distinguishes four types of action research, and explains their similarities and differences. She investigates the different choices researchers are facing during their research. Firstly, this concerns the component of change in the research, both the goal of the research (emancipation, empowerment, democracy) and the question as to what extent the change process itself is or should be part of the research. Secondly, she focuses on the theme of the relationship between researcher and researched party/co-researchers and the chosen methods, which imply certain relations between these two parties. Finally, she elaborates on a much debated point in action research: the question of reliability and validity or, in terms of action research researchers, trustworthiness, quality and validity.

We conclude with a chapter that wraps up the main findings of the contributions. We try to grasp the most pressing questions the reader – (future) researchers in action – might have, and give an indication of the answers, in view of the experiences and insights the authors have shared with us in their contributions.

We first ask ourselves: What does research in action offer? Without revealing too many of the answers you will find in the contributions, we can generally state that to some it is a useful dynamic research strategy in addition to other kinds of research, while for others, it forms a new paradigm that inspires them to a different scientific research approach. At the very least, it offers a framework for research that generates more and 'new' kinds of knowledge and an advanced evaluative framework to assess the results of these various kinds of knowledge.

Secondly, we want to know: Who is research in action intended for, with whom is the research undertaken and who owns the outcomes of the research? And on top of that, we want to know what purpose it fulfils. We think that research in action has various stakeholder groups. Besides the primary group of stakeholders there may be more supportive stakeholders, i.e. groups of people that need to be involved in the research to reach the ultimate goal but who themselves are not the prime subjects of the research. With emancipation, democratisation and empowerment as the flagships of research in action, it is important to maintain a clear view of who are the actual primary stakeholders.

Even with the goals and primary stakeholder group in clear focus, an important question remains: How can we bring about innovation and social change by doing research? To a large extent, the answer lies with the form research in action takes. Under ideal circumstances – which cannot be guaranteed beforehand and at all times – the researcher and the researched/co-researchers enter into a partnership which forms the basis for a mutual knowledge creation process in which individual and collective learning trajectories are linked to each other in the actual deed of doing research. Such an ideal learning environment can be only be created when there is a common ground of trust between the researcher and the researched/co-researchers, where every participant is able and allowed to take on a reflexive attitude, and where interaction between relevant actors is possible on various levels.

Finally, we ask ourselves why we need this new kind of research. The answer to this question is that the challenges our contemporary society faces are highly complex issues, involving a variety of interrelated aspects that cannot be tackled from just one disciplinary perspective. Interdisciplinary science is about designing and facilitating research processes that are aimed at a combination of examining, learning, development and change in organisations or communities. Consequently, a new type of researcher is needed. Beside the research skills they are traditionally equipped with, academics nowadays are required to be able to learn within a dynamic process and integrate the different kinds of knowledge that are being generated by various stakeholders in the research process. This implies not only empirical knowledge about 'matters of fact', but also normative claims about which values are worth striving for, and insight into power relations that have to be dealt with in order to be able to realise the strategic action plans with regard to the issues at stake. These are all the more important since within interdisciplinary projects, researchers are confronted with a plurality of evaluative frameworks that are being used by the different groups of participants. Moreover they have to try and transfer and apply the results of the knowledge creation process in a way that fits the actual situation. In sum, what we need are researchers with the competencies to build bridges between science and society. Scientists who realise that they cannot claim to possess the exclusive though all-inclusive knowledge to find *the* solutions for the complex questions, but who know how to convey scientific knowledge, with all its subtle nuances and uncertainties both to policy makers and corporate managers as well as to the general public.

Finally, we would like to register our awareness that we are researchers from the Netherlands. This means we have a western bias that is hard to avoid. We tried to bring in cases from other parts of the world, but these are limited in number. For a book that aims to be of particular use to PhD and Master's students who perform their research in a diversity of situations and under a variety of conditions, this forms a serious limitation. We therefore aim for a continuous dialogue with the readers of this book and others involved in research in action to broaden and sharpen our view and maybe repair the existing bias in a subsequent volume...

References

Greenwood, D.J. and M. Levin, 2007. Introduction to action research. Social research for social change. Sage Publications, London, 301 pp.

Reason, P. and H. Bradbury, 2001. Handbook of action research. Participative inquiry & practice. Sage, London, 468 pp.

't Hart, H., J. van Dijk, M. de Goede, W. Jansen and J. Teunissen, 1996. Onderzoeksmethoden. Boom, Meppel.

Valkenburg, B. and J. Lind, 2002. Reflexivity in international research on social policies. In: Van Berkel, R. and Horneman-Møller, I. (eds) Active social policies in the EU: Inclusion through participation? The Policy Press, Bristol, pp. 179-197.

Part 1: Research reports

2. Relating knowledge to policy making

Balancing environment & health-data, risks and actors in Flanders, Belgium

Hans Keune, Dominique Aerts, Ludwine Casteleyn, Dirk Wildemeersch and Lieve Goorden

2.1 Introduction

The Flemish Centre for Environment and Health investigates the relationship between environmental pollution and human health by measuring pollutants and health effects in (nearly 4,500) Flemish inhabitants for the interpretation of biomonitoring results for policy action. This chapter describes the development of an action plan by medical, environmental and social experts and policy representatives for policy interpretation of such human biomonitoring data We will focus in particular on the planning process and the practical research aspects of this endeavour, and pay particular attention to the role of the social scientists

Complex relationship environment & health on the societal agenda

At the international level experts increasingly agree that environmental problems have serious consequences for public health (McCally, 2002; Harremoës *et al.*, 2002). The number of environmentally related cases of cancers is rising, an increasing number of people suffer from respiratory diseases, fertility problems are on the rise, and so on. However, the (causal) relationship between environment and health is very complex and to a large extent unknown. Gaps in knowledge exist on the effects of small doses over a longer period and on the combined effects of different substances. In many cases it is very difficult to prove unambiguously that a relationship exists between environmental pollution and specific health effects (Ravetz, 2002; McCally, 2002; Harremoës *et al.*, 2002). Furthermore, scientists can and do differ in opinion on these issues. The question as to how to organise research and policy on these topics is difficult to answer because of the complexity of the matter. Our ability to measure even minute quantities of pollutants in human media by far exceeds our in-depth understanding of the public health meaning of the measurements. The complexity, however, relates not only to the relationship between environment and health, but also to societal choices. In addition to toxicity of agents, different societal perspectives and interests play an important role.

Centre for Environment and Health

The Centre for Environment and Health is funded by and works directly for the Flemish government. It was established in 2001, together with eleven similar Centres for policy relevant research. Their main task is scientific research on priority issues for government policy. A steering group, including representatives of governmental institutions, is attached to each Centre. In the steering group, policymakers discuss the knowledge production and valorisation with the researchers.

In the Centre for Environment and Health, environmental health experts from all Flemish universities, two non-academic research institutes and the Dutch University of Maastricht jointly investigate the complex relationship between environment and health. In addition, a social scientific expert unit focuses on risk communication, risk perception, and on processes of knowledge production, interpretation, deliberation and cooperation between different scientific disciplines and other social actors.

Human biomonitoring

The core activity of the Centre for Environment and Health between 2001 and 2007 was related to human biomonitoring. In a survey programme some selected pollutants and certain health effects were measured in the Flemish population. The focus was on three different target groups, each investigated in a separate campaign: newborn babies, adolescents and adults. Each campaign was carried out in eight areas of Flanders. These areas have different environmental characteristics, such as industrialised, rural (the countryside), urbanised, near waste incinerators, and near fruit orchards. One of the objectives of this biomonitoring was to focus on a comparison of exposure and health effects associated with these different types of environmental situation.

It is said that our ability to generate human biomonitoring data has exceeded our in-depth understanding of what the data mean to public health (human biomonitoring for environmental chemicals, national research council). One of the big challenges of human biomonitoring is therefore the interpretation, communication and further use in policymaking of the individual and collective results of such human biomonitoring campaign, answering the question: what should be done with the vast amount of data? A working group of scientists, governmental experts and policy representatives, most of them involved in the work of the Centre for Environment and Health, thus decided to prepare an action plan for the interpretation and use for policy of the human biomonitoring data.

2.2 Methodological approach

Action research: learning by doing

The procedure that emerged during the planning process under discussion may be qualified as a form of action research (Boog, 2002; Boog *et al.*, 2001; Coenen, 1987): direct intervention into practice is part of the process; the research is action-oriented. This ensures, to some extent, practice-relevance. This also means that complexities present in real life affect the development of the research procedure. Moreover, researcher and research subjects work together, instead of separately, without any hierarchical relationship and with interaction and participation as central concepts. From the perspective of the social scientists in the Flemish Centre for Environment and Health this research constitutes a two-layered interactive process. Firstly, the cooperation of the social scientists with other actors in the Centre: the different scientific disciplines and the policy representatives. Secondly, one of the goals is to involve actors (for example, stakeholders and local residents) external to the Centre. The social scientists support this process.

Reported research from the Pastille Consortium (2002) tells us that no best practices on dealing with action research are available. Researchers and research subjects should choose the approach they think is best for the research context. The Pastille Consortium points out that in action research a number of factors influence the interaction between scientists and representatives from policymaking. The extent to which policymakers are familiar with social sciences, by education or experience, will largely influence the interaction. The pressure for results is yet another factor. This is valid for both researchers and policymakers. The Pastille Consortium emphasises the need for policy relevant knowledge to be practical. This will, however, never be 'objective' knowledge because the researchers take part in the (hermeneutic) knowledge creation process. A main quality criterion for interactive research is knowledge based on the input of a diversity of actors.

The main result of the research process within the Pastille Consortium was the mutual learning by the participants. Similarly a process of joint learning by doing in which a diversity of actors contribute from their own expertise, views and background, is perhaps the best way to describe the practical approach within the framework of the Centre for Environment and Health.

The involvement of a variety of actors in the process as well as the use of different research and deliberative methods is relevant because of both scientific and social complexities. The concept of triangulation is helpful in framing such a joint learning approach. Denzin (1978) distinguishes four types of triangulation: (a) data triangulation, where data are collected at different times or from different sources, (b) investigator triangulation, where different researchers independently collect data on the same phenomenon and compare results, (c) methodological triangulation, where multiple methods of data collection are used, and (d) theory triangulation, where different theories are used to interpret a set of data. Janesick (2000) adds a fifth type to this list: (e) interdisciplinary triangulation, where the research process is informed not only by a single academic discipline, but by several disciplines. Involving different perspectives can also mean involving non-scientific actors. We may call this (f) transdisciplinary triangulation.

2.3 Results: practice

Phased evaluation

Upon the initiative of the Environment Department of the government, medical and environmental scientific experts and policymakers, together with social scientists embarked on the preparation of an action plan for interpretation and policy measures with regard to the human biomonitoring results (Koppen *et al.*, 2005). In the beginning discussions in the working group focussed mainly on interpretation of biomonitoring data from the perspective of environment and health sciences. Consultation of scientific experts as well as desk research was considered to provide the necessary knowledge and answers. Later on in the conceptual process other elements were introduced: complementary assessment criteria, complementary assessment methods and involvement of other actors in the process. This generated a level of complexity in the process which asked for systematic approach. We first describe the analytical steps that were proposed. Further on in this chapter we present the complementary issues, methods and actors as they were integrated by means of a practice cycle.

The working group defined that the biomonitoring results were to be assessed for different aspects in three successive analytical phases. The first phase would focus on the question: how are specific results to be seen with regard to public health risks? To a large extent in this phase the discussion was to focus on the availability and use of reference values for interpreting the data. This is quite problematic since knowledge on these issues is still rather limited. Only with regard to lead there exists a 'norm' to which levels of lead may be compared in order to judge whether there is reason to be concerned about human health risks. Therefore the average range found for each pollutant or health effect is used to assess which biomonitoring results are relatively higher. A comparison is also made with research outcomes from other studies, e.g. from abroad. The second phase would address the question: what are the causes of a specific monitoring result? For example, causes may be environment-related or lifestyle-related. In the third and final phase the key question would be: can we identify a (local) source for the pollution that can still be influenced by policy measures?

Above we outlined the fact that the development of the action plan was initially thought of as a merely scientific quest. It was anticipated that with the right group of experts the interpretation with regard to policy priorities would follow automatically. However, while trying to build bridges towards policy interpretation, the limitations of an exclusively scientific endeavour clearly showed. Social scientists introduced issues of economics (cost-benefit; cost-efficiency), social preferences, feasibility of policy measures. This raised difficult questions, e.g. on policy priorities when factors other than (medical, public health and environmental) scientific ones also had to be taken into account. Given the many uncertainties and limited specialised knowledge, no scientist or group of scientists dared lay claim to possessing the necessary and overarching knowledge for answering difficult questions which were introduced by the social scientists. The social scientists proposed the formation of a jury (potentially including representatives of local societal groups as well as individual citizens) to judge relevant data and knowledge in order to give advice to the government. Furthermore, they reckoned the development of a stepwise procedure 'from data-interpretation to decision making' was critical to arrive at a functional action plan.

Practice cycle

The social scientists developed what they called 'a practice cycle' with different procedural steps, actors and roles in order to move from scientific assessment to a process of decision making. For each of the three analytical phases, the social scientist defined the main questions to be answered as input to the development of this practice cycle. These questions were:

- Who is responsible for further research based on the human biomonitoring results?
- Who should steer the process?
- Whose knowledge or opinion is relevant?
- Who should decide on policy options?
- Who should be informed about the developments and outcomes during the process?
- What are the key points of interest during the process?

The social scientists investigated the views among the participants of the working group by means of a short questionnaire. On the basis of the outcomes of the questionnaire, it would be possible to propose practice cycles, i.e. proposals for organising the process and for involving different kinds of knowledge and actors. However, this step was not unproblematic. The first questionnaire resulted in very few responses. The need for such an approach was not recognised by the other members of the working group, nor did such a focus on the issues dealt with in the questionnaire seem urgent to them. This can partly be explained by discomfort in the face of social scientific expertise and methods. This is often associated with a vast underestimation of the complexity of social scientific expertise. As time went on however, the issues dealt with in the questionnaire kept on reappearing in the discussions as essential issues to be dealt with in the action plan, albeit in an unstructured and unsatisfactory manner. As a consequence the working group members became increasingly convinced of the need for a more structured approach. The second questionnaire attempt yielded more response and resulted in much food for further thought and practical application in the action plan.

The advantages of this 'internal reflection' by questionnaire were several. It gave structure to the process of brainstorming about these complex issues. In itself there is of course nothing wrong with unstructured brainstorming, but at some point there is need for sound analysis and synthesis. At the same time the collection of views and ideas on an individual level through a questionnaire enriched the discussions with diversity from different perspectives. In group brainstorming this diversity may 'disappear' unnoticed and partly remain in the dark. This diminishes opportunities for the actors participating in the brainstorming to learn from other perspectives. From the perspective of the social scientist this offered a good opportunity, not only to collect material for the issues at stake, but also to get a grip on a process that does not necessarily 'belong' to the social scientific field. In other words, it provided good opportunities to analyse the process from a social scientific perspective with a totally different 'set of reflexes'.

An interview round with regard to another issue in the framework of the Centre for Environment and Health (Keune and Goorden, 2004) revealed that both scientists and policymakers indeed feel a need for a kind of structured reflection on their work. In everyday practice however, no time or priority is reserved for such an exercise. The view of most of the working group members to this recognised benefit of reflection differs from the actual response in daily practice to such social scientific approach. Appreciation for the results from the internal reflection did not necessarily result in a change of practice (i.e. reserving time for reflection) and attitude. Under pressure of the working load, routines prevailed and openness to these 'new ideas and approaches' disappeared. This meant that at later stages in the process both scientists and policymakers had to kind of restore the more open view towards other, social scientific, perspectives and methods. This leads to the conclusion that in complex processes on complex social issues it is not enough to integrate social scientific perspectives and methods. The actual involvement of social scientists throughout a cooperative process is also essential to guarantee continuity and social scientific quality.

One example of contribution from questionnaire-based reflection was a more precise definition of the main questions dealt with in the action plan, especially with regard to the first phase. The central question of the first phase became: what relative priority do the different biomonitoring results have for policy action? For example, are levels of heavy metals in a certain area more important with regard to public health than levels of pesticides in another region? The outcome needed was a ranking of biomonitoring results for policy.

The practice cycle that was developed based on the information collected through the internal reflection, is made up of cyclical steps, to be taken during each phase of the action plan. It includes (1) deciding how to operate and which actors to involve during the process, (2) desk research on the biomonitoring results and (3) expert consultation. Then it includes (4) bringing a synthesis of the desk research and expert consultation before a jury of stakeholders. And (5) a synthesis of desk research, expert consultation and jury advice is presented to the administration. And finally, (6) the government decides on the next steps. During all of these steps, external communication about the process was felt to be important and is therefore included.

Another outcome of the internal reflective questionnaire was a reservoir of potentially relevant actors for different steps during the practice cycle. This varied from scientific experts to be consulted for scientific interpretations, to stakeholders for input on societal issues, to the broader public for communication purposes. Furthermore, the relevance of taking into account social, economical and policy issues was recognised.

For the expert consultation the social scientists proposed the Delphi method (e.g. Slocum, 2003). One drawback of this kind of approach is that it is rather time consuming: experts have to be consulted at least twice, first for individual assessment based on information regarding the problem, secondly for individual assessment on the different inputs from the total group of experts. Because of practical considerations (time and complexity of the process) it was decided to only organise one round of consultation. This example highlights a specific aspect of such inter- and transdisciplinary cooperation: negotiation. It is often impossible to fully implement all idealistically relevant aspects of a methodology in practice. Not only because of practicalities of work in real daily life, but also because people with different background, visions, experiences have to find a common understanding of what will be good research or action practice.

Jury and multi-criteria analyses

The fact that no scientist or group of scientists dared claim to possess the necessary and overarching knowledge for answering all difficult questions, prompted the social scientists to propose working with a jury. The jury will be formed by experts, stakeholders and (other) citizens. The reservoir of potentially relevant actors will form the basis for selecting candidates.

For the jury in the first phase of the action plan, a multi-criteria analysis was developed. The three main (groups of) criteria are: seriousness of environmental & health risks, feasibility of policy measures and societal aspects. The content of these three criteria will be articulated during the second step of the practice cycle: desk research and expert consultation. In the multi-criteria analysis the jury for each biomonitoring result discusses the relative importance (weight) of each criterion. The outcome will be more than merely a numeric ranking of policy priorities though. The jury process will be a group discussion in which participants can learn from each other and where views, arguments and concerns will be exchanged and monitored. This meant that the multi-criteria analysis functions as a method for structuring discussions and for supporting reflection during the deliberations. Transparency and practical employability are therefore essential features.

On the basis of its deliberations, the jury is expected to formulate a policy recommendation for the government. Of course, the government may decide otherwise. But part of the process is the commitment to transparency: once the government decides, it communicates not only the decisions, but also the arguments for it. The government also promises to respond to the suggestions, arguments and concerns raised during the process that led to the jury recommendation.

Action plan in policy practice

The action plan was accepted by both the Centre for Environment and Health and policy representatives, and was adopted by the government. A pilot project that deals with the results of the first campaign of biomonitoring (newborn babies) started in 2006. It is impossible as yet to predict how this action plan will work in practice. The fact that the government is prepared to test the procedure in practice is promising though. Moreover, the Flemish experience will be the inspiration for a European Union pilot project with regard to human biomonitoring.

2.4 Conclusions

Boundary work

The complex interaction, cooperation and negotiation between different scientific disciplines and between scientists and non-scientific actors such as policymakers can be seen as a form of 'boundary work' (Gieryn, 1983; Jasanoff, 1990; Hoppe, 2002). The social scientists contributed in structuring this complex interaction. In Hoppe's (2002) words, a dialogical, pragmatic model interaction between experts and policy representatives was established during the development of the action-plan. Most of the policy representatives and policy experts involved from the start were medical, public health, technical or environmental specialists by training, as are most of the scientists involved in the Centre for Environment and Health. In that sense there was no major divergence in the way the issues under discussion were approached. Building the bridge towards political interpretation nevertheless proved problematic: an obvious discrepancy in issue framing between experts and policy makers emerged, to a large extent due to uncertainties regarding the results and the need for quick policy actions. In order to bridge this gap the involvement of social scientists proved to be helpful. A reflective contribution paved the way for the participation of a diversity of actors, enriching the assessment with other than technical medical, public health and environmental criteria. Moreover, clear procedures for cooperation, deliberation and decision making were introduced.

Action research

The effort sketched in this paper, the development of the conceptual action plan, turned out to be a rather labour-intensive exercise. In the activities of the Centre for Environment and Health, the time available for reflection on the work in progress is rather limited. Moreover, most actors involved in the process do have overloaded agendas. Discussing complex issues takes time and energy, often goes hand in hand with a lot of 'paperwork' and does not always result in consensus. Part of the work also deals with unforeseen complexities, adding to the time investment. There is always a tension between the pace of nuanced, scientific discussion and treatment of such issues on the one hand, and the pace of day-to-day politics on the other. This is the eternal split scientists are forced into by political authorities. The research product must be of the highest scientific standard and able to survive all possible criticism even on a very specific, nuanced level. At the same time the product must be delivered quickly and should be simple. A 'delay' of several years between the start of a survey and the political actions based on the research results is difficult to explain to the press, to the public as well as to the politicians themselves. This 'tension' needs attention, for example in the way policy-relevant research is communicated.

Quite a lot of time is spent on actors getting used to working together. Trust-building takes time and effort and is an important part of the challenge. Actors have to find new roles for themselves to some extent. In the case of the Centre for Environment and Health scientists were suddenly forced to discuss their work with experts from (sometimes totally) different disciplines. They also had to talk to politicians about politics. Government representatives suddenly had to discuss science and different fields of policy expertise (the Ministries of Environment and Health) also had to come to terms with each other. Apart from role-seeking, this also requires (new) procedures. Furthermore, common vocabularies need to be developed: different scientific disciplines, policy makers and other actors use different (technical) language, have different cultural backgrounds and a different knowledge base.

Social science

The daily practice of the Centre for Environment and Health clearly differs from a laboratory situation. In a laboratory research conditions are controlled (or controllable) to a large extent and complexities are restricted to a certain degree. The research practice of a social scientist in a setting such as the Centre is something completely different. In the laboratory atoms and molecules more or less do what they are expected to. In social practice research subjects will go into discussion with the researcher, they talk back (Bal *et al.*, 2002). We can therefore speak of a 'stubborn' practice.

The appreciation of and openness towards other views, other disciplines, and transdisciplinary knowledge seems to have increased in the Centre. This runs virtually parallel with the development of the work of the social scientists. Fisher (2000) uses the concept of coordination when describing the role of social scientists. He speaks of the necessity of developing innovative methods for coordinating between different discursive processes and institutions.

This role of the social scientists within the framework of the Centre for Environment and Health was to some extent part of the official call from the governmental authorities for the establishment of the Centre, and was also present in the research agenda of the social scientists as agreed upon by the other scientific disciplines at the start. Part of the approaches used by the social scientists was also included in the guidelines for external communication of the Centre developed in close cooperation with both the scientists and policymakers involved in the work of the Centre during the first year (Goorden *et al.*, 2002).

At the 'Ten Commandments level' (at the abstract level of ambitions, like in the bible) the main principles for external risk communication are relatively easily understood and agreed by the actors involved. In practice though, a lot of practical problems that do occur show that the complexity is sometimes enormous and underestimated. Practice is therefore to some extent a test for the appearance of consensus on a 'Ten Commandments level': differences in vision, interpretation and preference will come forward. Furthermore, we have to take into account the sometimes rather large discrepancy between the social scientific perspective on issues like risk communication and the way other scientific disciplines or policy representatives perceive it. Social scientists have other knowledge and experience from their disciplinary background.

Thus, if we look at the developments in real-life practice, the role of the social scientist can best be characterised as 'emancipating'. In the beginning the role of social scientists according to some of the other actors within the framework of the Centre was more or less perceived as limited to communication issues: work to be done after the scientific conclusions were drawn. Gradually other scientific disciplines as well as policy makers showed more appreciation for 'other' capabilities and forms of knowledge: from being complicated knowledge from 'another planet', the contribution of social scientific insights and methods was gradually taken more seriously as a necessary asset for the complex scientific and policy endeavour.

Acknowledgments

The study was commissioned, financed and steered by the Ministry of the Flemish Community (Department of Science, Department of Public Health and Department of Environment). The work was performed by the Centre for Environment and Health (promoters: Prof. Willy Baeyens, Prof. Nik Van Larebeke, Prof. Greet Schoeters, Dr. Vera Nelen, Prof. Ilse Loots, Prof. Stefaan Dehenauw, Prof. Geert Molenberghs). We gratefully acknowledge the collaboration of the (nearly 4,500) participants in the biomonitoring. Field work was done by trained nurses of the Antwerp Provincial Institute of Hygiene: Liliane Thys, Els Goossens, Nancy Maes, Ghis Meyssen, Babs Van Nesselrooij, coordinated by dr. V. Nelen and E. Van de Mieroop.

References

Bal, R., W.E. Bijker and R. Hendriks 2002. Paradox of scientific authority - about the societal influence of advice of the Health Council. Health Council, The Hague.

Boog, B., 2002. Action research: An update. Sociale Interventie 11 (4): 27-40.

Boog, B., H. Coenen and L. Keune (eds.), 2001. Action research: Empowerment and reflection. Tilburg University Press, Tilburg.

Coenen, H., 1987. Action research as exemplary learning. A contribution to the foundation of the methodology of action research. Utrecht University Press, Utrecht.

Denzin, N.K., 1970. The research act in sociology. A theoretical introduction to sociological methods. Aldine Publishing Company, Chicago.

Fisher, F., 2000. Citizens, experts, and the environment, the politics of local knowledge. Duke University Press, Durham, London.

Gieryn, T., 1983. Boundary-work and the demarcation of science from non-science: Strains and interests in the professional ideologies and cultures. American Sociological Review 48: 781-795.

Goorden, L., H. Keune, I. Loots and R. Mertens, 2002. Guidelines for external communication. Centre for Environment and Health.

Harremoës, P., D. Gee, M. MacGarvin, A. Stirling, J. Keys, B. Wynne and S. Guedes Vaz (eds.), 2002. The precautionary principle in the 20th century, late lessons from early warnings. Earthscan Publications Ltd, London.

Hoppe, R., 2002. Rethinking the puzzles of the science-policy nexus: Boundary traffic, boundary work and the mutual transgression between STS and Policy Studies. Paper prepared for the EASST 2002 Conference, 'Responsibility under Uncertainty', York, 31 July-3 August 2002.

Jasanoff, S., 1990. The fifth branch: science advisors as policymakers. Harvard University Press, Cambridge, Massachusetts, London.

Janesick, V.J., 2000. The choreography of qualitative research design. Minuets, improvisations, and crystallization. In: Denzin and Lincoln, Handbook of qualitative research. Thousand Oaks, California.

Keune, H. and L. Goorden, 2004. Accidental complexity, societal embedding of science on health & Environment, case study Genetic Research. Centre for Environment and Health.

Koppen, G., H. Keune and L. Casteleyn, 2005. Action-plan biomonitoring results. Centre for Environment and Health.

McCally, M. (ed.), 2002. Life support, the environment and human health. MIT Press, Cambridge, Massachusetts, London.

Pastille Consortium, 2002. Indicators into action: local sustainability indicators sets in their context. 5th Framework Programme, European Union.

Ravetz, J., 2002. The post-normal science of precaution, paper for the thematic workshop of the joint research centre 'managing uncertainty in science for sustainability: future research challenges for Europe', at the Launch Conference of the 6th Framework Programme: 'European Research 2002, The European Research Area and the Framework Programme', Brussels, 11-12-13 November 2002.

Slocum, N., 2003. Participatory methods toolkit, a practitioner's manual. viWTA and King Baudouin Foundation, Belgium.

3. Innovation through action

An action research journey with smallholder farmers in Limpopo Province, South Africa: experiences of soil fertility management

Johannes M. Ramaru, Jürgen Hagmann, Zacharia M. Mamabolo and Michael H. Netshivhodza

3.1 Background

This chapter documents the learning processes within the framework of innovation of soil fertility management practices that emerged from the implementation of Participatory Extension Approach (PEA) as part of service delivery reorientation within the Limpopo Department of Agriculture in South Africa. Other major innovation processes were developed around the same time and they include soil and water conservation, small-scale seed production and livestock management. All these innovations were developed by teams of researchers, extension officers and contributions from the farmers following the guiding principles of the PEA process. The author chose to use a soil fertility management case to illustrate the learning curves that researchers underwent together with farmers primarily because he was part of the learning team. Although reference is made to other villages in the later years of implementation of soil fertility management, the description of the learning process is rooted more in the first two seasons and in particular in three villages of the Capricorn districts, namely GaThaba, GaMogano and Spitzkop. In short, the chapter gives a narrative description of what transpired during the interaction between researchers, extension officers and farmers, the processes involved, the lessons and conclusion.

The processes took place in the context of the attempt to increase the efficiency of service provision to rural farmers by Limpopo Department of Agriculture (LDA). Limpopo Province is one of the nine provinces in South Africa with most of its people living in the rural areas. With 85% of the people involved in farming, agriculture is one of the pillars contributing to the economy (Anonymous, 1995; Schuh, 1999). After the establishment of the new government in 1994, it was realised that the majority of smallholder farmers in rural areas could not access adequate services from the government because existing mechanisms were largely modelled on commercial farming – the notion of 'one size fits all'. This prompted the South African government to develop policy statements[2] (Shao *et al.*, 2004) aimed at addressing poverty among the majority of rural people. In Limpopo this led to the establishment of a programme in 1998 called Broadening Agricultural Services

[2] Mission statement of the National Department of Agriculture is to "Ensure equitable access to agriculture and promote the contribution of agriculture to the development of all communities, society at large and the national economy, in order to enhance income, food security, employment and quality of life in a sustainable manner" (Anonymous, 1995).

and Extension Delivery (BASED) which was implemented through the funding support of German Technical Cooperation (GTZ).

3.2 Initiating service delivery reforms through the use of PEA

The starting point for the BASED programme was to conceptualise the strategy for service delivery. Within this strategy, the first task of the BASED programme was to develop a participatory and learning approach to development. This strategy was systemically integrated with other learning processes in the programme, namely competency development, case study development, innovation and service development and approach mainstreaming. The starting point, the Participatory Extension Approach, had previously been developed in Zimbabwe and was perceived to be very effective. This approach is described by Moyo and Hagmann (2000) as a process that enhances self-organisation at the community level and promotes a demand-oriented extension service system. In addition, Hagmann *et al.* (1998) explain that Participatory Extension Approach is a learning framework which integrates different methodologies and tools (e.g. Participatory Rural Appraisal [PRA], Farmer Field Schools [FFS] and Participatory Technology Development [PTD]) in a rigorous learning process to deal with different topics in rural development. To prepare for the operationalisation of the processes, the initial team of 40 extension officers and scientists from Capricorn and Vhembe districts in Limpopo were exposed to some of the different competency development process steps (see Table 1).

Reflexive processes from these initial learning experiences provided insights that lead to the development of the PEA (see Figure 1) within the context of Limpopo province. This kind of learning cycle as described by Allen *et al.* (2002) and Marais *et al.* (1997) is a basis upon which different stakeholders would follow a series of cycles in which they plan activities, act and reflect on the change process. Similarly, experiences and lessons from the implementation of PEA were derived from a series of assessment processes and fed back to various levels in the communities and within the ministry of agriculture. The modified PEA learning cycle was found to be composed of six main phases, namely initiating change; searching for new ways; planning and strengthening local organisational capacity; experimenting while implementing actions, sharing of experiences and reflection on lessons learned and replanning.

3.3 Developing processes within a framework soil fertility management

Using the skills acquired from the training workshops on PEA, extension officers and scientist started to develop and use different tools and codes for awareness creation. It was from the facilitation of needs during community meetings that the problem of declining soil fertility was identified by farmers from both Capricorn and Vhembe districts. So serious was the problem that one of the farmers during the meeting shared the following sentiments '*...we are adding fertilisers but yields are not increasing. Can't you scientists recommend to us a*

Table 1. Summary of PEA process and activities undertaken by the 40 extension officers and scientist during 1998-2000.

Period	Action steps
August - September 1998	Familiarisation of the approach to the officers and communities
October 1998 - February 1999	Household surveys and local institutional analysis
March 1999	Exposure visit by farmers and LDA officers to Zimbabwe and feedback about the visit
April - June 1999	Introduction of the concept of Umbrella Organisation (UO) and strengthening of local organisation
August 1999 - February 2000	Analysis of the farmers technical problems and introduction to experimentation processes involving farmers
	Using soil fertility management (SFM) to learn to understand the process of innovation
	Develop other innovations in the area of soil and water conservation (SWC), small-scale seed production (SSSP) and livestock management (LM)
March - April 2000	Introducing the concept of mid-season evaluation (MSE) as a tool for farmer to farmer sharing
July - November 2000	Process review and re-planning of activities for the next season

fertiliser of another colour – we mean a new type of fertiliser?' The soil scientists present during the meeting thought they could help but did not know how. The initial opinion at the time was to encourage farmers to apply fertilisers at least of a different kind. Neither the scientists nor the farmers knew that this was the start of the learning journey.

To get prepared, the research team developed an operational plan based on the participatory technology development (PTD) steps as described by Van Veldhuizen *et al.* (1997). The plan was designed to answer the following questions: what steps would be taken to solve the problem of declining soil fertility, what would be the objective/s of each step, what activities would be carried out for each step, which methods or tools would be used, when will activities be executed, who will be partners in the implementation process, what materials and resources will be needed. The development of these plans by the scientist and extension officers has helped to shape a shared vision of how the process was likely to unfold.

The forthcoming sections share experiences captured largely during the first year of finding an alternative solution to the problem faced by farmers of declining soil fertility management in three villages for three cropping seasons.

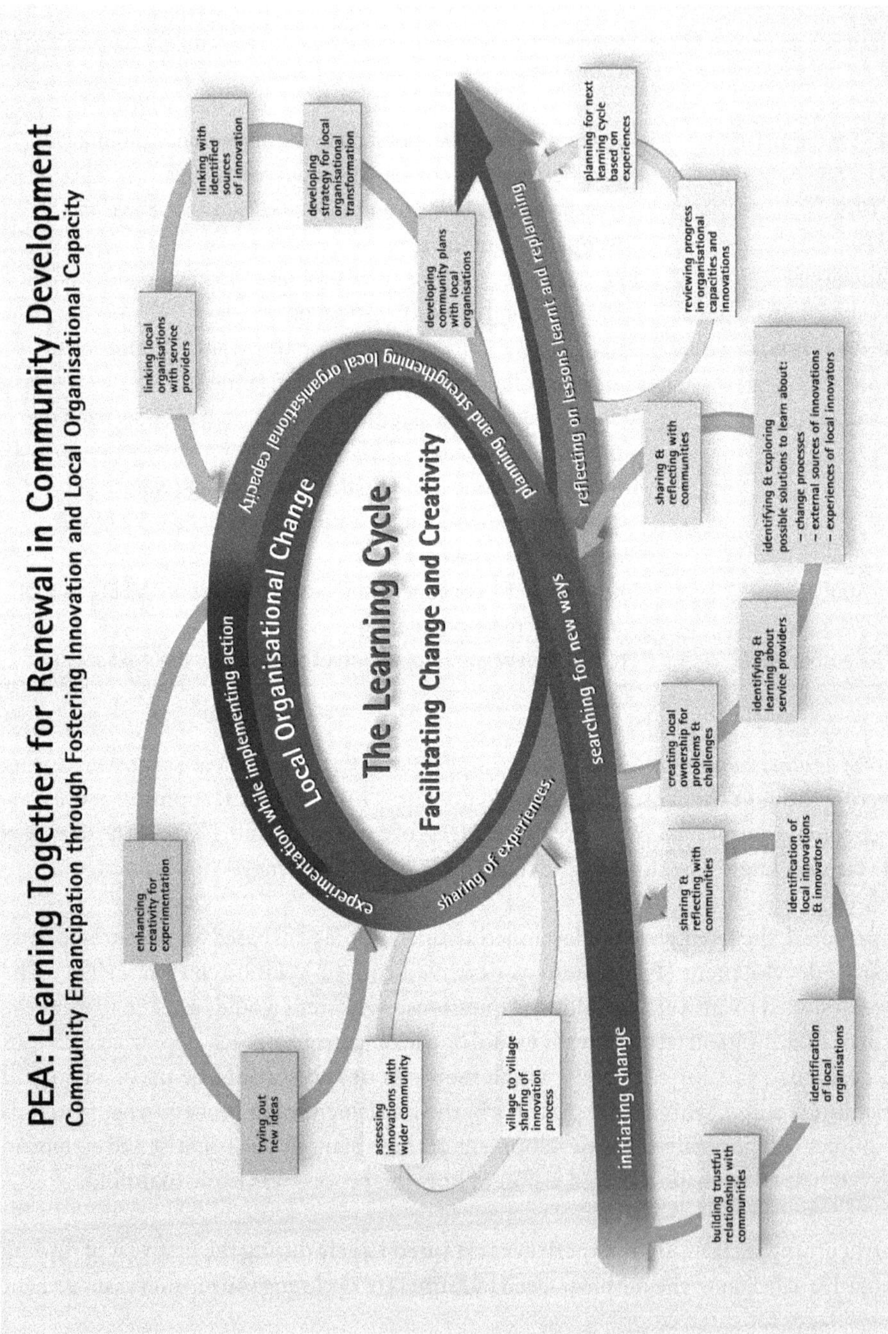

Figure 1. The basic phases of the learning cycle as developed through an action-research process in South Africa (Hagmann et al., 2003).

Perceptions of farmers about the declining soil fertility

The research team (composed of the scientist and extension officers) wanted first to assess farmers' perceptions of soil fertility in the area. This was seen as an opportunity to learn about practices previously used by farmers and get suggestions from them on how to move forward. At this stage, members of the research team were trying to bring into practice what they had learned from PEA and the PTD workshops – how to involve farmers right from the start of participatory research processes. Key questions that guided the discussions during the meeting with farmers were:

- What do you understand by declining soil fertility?
- How does this problem present and how long has it existed?
- What did you do to address the problem and what were the results?
- What do you think should be done to solve this problem in the future?

These questions provided an opportunity to start discussions with farmers and were followed by deeper discussions to unpack some issues. Table 2 shows some of the responses from farmers in the problem analysis exercise.

> 'Just like when we go to the doctor for a check up when we are sick, we now know that soils also need to be tested or doctored to determine their nutritional status and for farmers to apply necessary inputs' (farmers from Spitzkop, March 2000).

During the discussions, farmers indicated that soil samples had been taken before by the extension officers and fertiliser agencies, but the results were never discussed with them. It was then agreed with farmers that since there were no previous records on soil analysis, be it from the extension officer or the fertiliser companies, soils samples would be taken from farmer groups that were interested. The research team saw this as an opportunity to get to know the nutritional status of the soils and widen the scope of options that could be tried out by farmers. Furthermore, agreement at the end of the meeting was that those who were part of these awareness meetings would hold meetings with their constituencies to report what transpired. During these meetings, they would also share with the members agreements on steps that would be followed to solve the problem of declining soil fertility. The farmers

Table 2. Perceptions held by farmers on declining soil fertility in one of the three villages.

Village	Description of the problem
GaMogano	Declining yields
	Crops become stunted after weeding
	Maize leaves turn yellow, even after application of fertilisers
	Maize grown on infertile soils is badly infested with striga

also agreed that they were going to discuss with their colleagues how they were going to pay for the cost of analysis since this service was no longer given for free.

Farmer mobilisation for paying for the costs of soil analysis

It emerged from several interactive meetings that although extension officers and fertiliser companies used to take soil samples from farmers' fields, farmers working in the project groups[3] were often unaware that their soils had been tested, as they were neither involved in the sampling nor fed back the results. Having been largely ignored by extension agents in the past, most of the farmers from communal groups[4] were also unaware of the benefits of soil analysis. The first awareness meeting aroused the interest of farmers from the communal field to have their soils tested as well.

Farmers agreed during their meetings that they were going to collect money to pay for the costs of soil analysis. There was a difference in the manner in which farmer groups paid for the costs of soil analysis. All project groups contributed money from their joint project savings account, whereas the communal groups had to make individual collections to make up the required amount. This meant an initiation to mobilise the communal farmers who had previously operated independently from the extension officers and researchers. Soil samples were analysed for physical and *chemicals properties* at a 60% discount, negotiated by the research team on behalf of the farmers (Table 3). Once farmers had collected all the money, arrangements were made with their representatives to set up a demonstration on how to take soil samples. The demonstrations to the farmers were done in the field by the research team, thereafter farmers had to do it themselves. When doing the soil sampling,

[3] Project group is a group of farmers working closely with the extension officer and following the orders and decisions of the officer. More often than not, the group constitution governing the farmers was developed by the officer.

[4] A communal group is a group of farmers who practise their agriculture independently without the help of the local extension officers. Their group is formed on an informal base and they are not obliged to form a management committee. The officers have no influence on the decision-making processes of the group.

Table 3. Soil sampling in the pilot and other villages (R 8=1 US$).

Factor	Trend over seasons			
	99/00	00/01	01/02	Total
No of villages	3	12	8	23
No. of samples collected	16	29	23	68
Money paid by farmers	R 1,070	R 1,885	R 2,065	R 4,990

farmers would argue and correct each other when mistakes were made, which was a sign that they were beginning to comprehend the soil sampling process.

Simplifying feedback of soil analysis results to the farmers

> 'We are used to seeing our extension officers and fertiliser agents taking soil samples from our fields, but we never hear from them again' (comments by farmers at GaMogano village - August 1999).

The last part of this statement by farmers from this village created a lot of discomfort to the research team. It meant that farmers were expecting us to do things differently from the way it was done by other researchers in the past, but most importantly, that they were curious to know the nutritional status of their soil.

After getting the results, the research team used guidelines from Borman *et al.* (1989) to rate the levels of essential elements and suggest different options for improvement. To simplify the presentation of the scientific information to farmers, the team developed several tools. They made colour posters illustrating the symptoms of nutrient deficiency and using the local language. They also used a lever scale (Figure 2) to illustrate the water-retaining capacity of soil and manure, and a magnet and pins to demonstrate the importance of manure in increasing the nutrient-retaining capacity of soils. These tools proved highly effective as a means of presenting analytical results in a simple format that enabled farmers to understand the nutritional status of their soils (see the comment from a PhD researcher in Box 1).

Figure 2. Level scale for demonstrating the water–retaining capacity of the soils.
'Now our extension officers are able to explain to us in a simplistic and practical manner things that we thought were complicated' (A farmer at GaMogano, August 1999).

Box 1. E-mail message reaction to Ramaru *et al.* (2000), Thursday, May 31, 2001 10:35 AM.

Sir, I was thrilled reading through a piece of work you wrote on improving soil fertility management in South Africa. I am a WARDA PhD research fellow in IITA Ibadan. My field of research is soil chemistry with emphasis on how to improve soil fertility through a participatory approach in a rice based soil in South-Western Nigeria. I know that information on soil testing result is valuable and needed for soil fertility management. All the while my major challenge was how to get such information across to farmers. However, on reading your publication I was glad. ...I would also appreciate any other information that would aid my research work... Thanks...

The first results were fed back by the researchers at a village called GaMogano and there was a frightening misunderstanding at the start of the meeting. The meeting opened with the usual prayer, welcome and introductions. Researchers outlined the objectives of the meeting and linked the meeting to those held in the past. Next, the research team had planned an exercise designed to raise awareness about declining soil fertility, and to enable farmers to understand how the causes of declining soil fertility lead to the results shown in the soil analysis. Suddenly, the mood of farmers changed when they were asked to form small groups to do the exercise. To everyone's amazement, they refused to do so, arguing that they had formed groups in the previous meeting and handed out information, and they were fed up with it. The farmers wanted to see the results of the soil analysis immediately. Although the facilitators tried to explain the purpose of the exercise and how it was linked to feedback on the results, farmers were adamant that they would only stay at the meeting if they were given the results of the soil samples first. The local village leaders tried to persuade them to keep to the planned programme, but when the farmers refused to back down, the facilitators aligned the programme according to the wishes of the farmers (Ramaru *et al.*, 2000).

Before soil results were presented, farmers shared their knowledge on the nutritional levels of soils that had been analysed. As an indicator of the level of soil fertility, they provided information about the plants that would be found on poor or fertile soils (Figure 3). Discussions on the results of the different sets of soils were held and, interestingly, the knowledge of farmers and laboratory results correlated well. The results of soil analysis showed a higher level of nutritional elements in the project fields than in the communal fields (Table 4). Based on the results, a set of options for soil fertility improvement were suggested by the research team and they included application of fertilisers, manures, compost, and dolomitic liming material.

Farmers were excited to learn that they had a choice from a pool of options to improve their soil fertility. They could also try out different combinations of the inputs and then

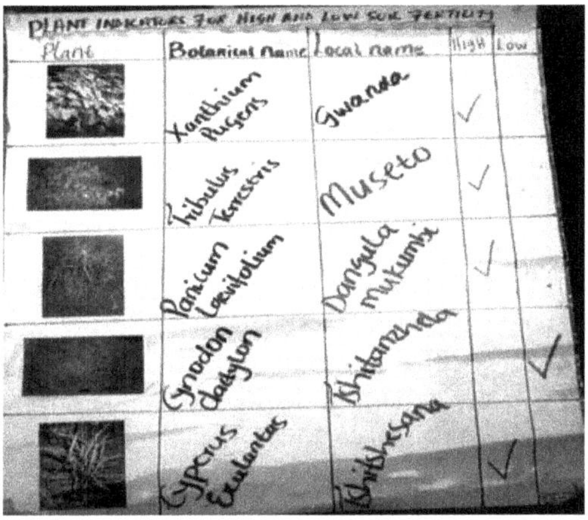

Figure 3. Illustrations by farmers on plants that are found on poor and fertile soils.

Table 4. Results of chemical soil analysis taken in 1999.

Soil parameter	GaThaba		GaMogano		Spitzkop	
	PF N=1	CF N=7	PF N= 2	CF N=2	PF N=1	CF N=3
pH (water)	5.6	6.3	5.8	7.2	6.0	5.9
P (Bray1) mg/kg	15.0	15.1	28.1	18.4	6.0	0.9
Ca ($cmol_c$/kg)	2.4	1.8	1.0	3.8	1.3	2.2
Mg ($cmol_c$/kg)	0.98	0.80	0.54	1.54	0.67	1.02
K ($cmol_c$/kg)	0.39	0.26	0.35	0.45	0.18	0.27
Na ($cmol_c$/kg)	0.10	0.05	0.05	0.06	0.05	0.06
Titratable acidity ($cmol_c$/kg)	0	0	0.08	0	0	0

Note: PF = project field; CF = communal field, N= number of samples.

share the lessons and the results. In the past, the decision on what to apply in the field came as a command from the extension officers. But at this stage there were still more questions than solutions. The first challenge that farmers had to address was how to access some of the options discussed.

'We made it, the tools worked according to plan'. This was an expression made by one of the officers during the reflection meeting.

Farmer linkages with the suppliers for inputs

> 'In the past it was only the extension officer who knew where to get inputs. When he/she died, he/she would be buried with the input, and when he/she was transferred, the input would also follow' (farmer in Spitzkop, 1999).

Having agreed with farmers to start improving their soil fertility status using various types of inputs, farmers were faced with challenges such as:

- How are we going to access different inputs and purchase them?
- How are we going to get contact with the various suppliers and the quotations for prices of inputs?
- What would be the most cost-efficient way of buying inputs? What about the delivery of the inputs to the communities?
- How could the newly established umbrella organisations[5] (UOs) help in facilitating the process of accessing the inputs?

The formulation of those challenges helped the research team and farmers to develop strategies for accessing suggested inputs.

Farmer representatives, mainly from the established umbrella organisations, were chosen to start the negotiation with the input suppliers. The meetings were such that more than one supplier would be invited to offer services and discuss prices and terms of delivery of the inputs to the community. Agreements would be reached about the volume of inputs bought by farmers, how they could qualify for a discount and how farmers would make payments. After meeting with the suppliers, farmer representatives would arrange community meetings together with the local leadership for feedback on the agreements with the service providers. The final decision to choose an appropriate input supplier is made during feedback meetings by representatives to the communities (Ramaru *et al.*, 2004). This process was used mainly for accessing fertilisers and dolomitic lime (see Figure 4). The difference with other inputs was that with cattle and poultry manure, the farmers only had to pay for hiring transport since the inputs were donated by big commercial livestock companies. The overall results from this process are shown in Table 5.

The original assumption by the research team was that farmers were not buying inputs such as fertilisers because they did not have money (mostly the communal farmers). This process revealed that once farmers were aware of the soil nutritional status, they were keen to form community forums such as umbrella organisations that would link with input suppliers. These processes made farmers internalise negotiation skills. They also started to exercise their purchasing power to their own benefit.

---

[5] Umbrella organizations (UOs) have just been formed. They are functioning as a village forum from which all the different farmer groups would select two representatives for the purposes of accessing services.

Figure 4. Lime delivered by the input supplier to the village.

Table 5. Number of villages and farmers involved in the acquisition of inputs (total amount of money that was contributed by farmers to purchase inputs, R 8 = 1 US$).

Input	Factor	1999/2000	2000/2001	2001/2002	Average in 3 years	Money collected by farmers [R]
Fertilisers	number of villages	3	4	26	11	
	number of farmers	494	333	2,211	1,012	
	number of bags (50 kg)	887	624	3,818	1,776	458,547
Dolomitic lime	number of villages	3	2	9	5	
	number of farmers	74	4	198	92	
	number of bags (50 kg)	648	18	2,541	1,069	39,149
Poultry manure	number of villages	3	4	2	3	
	number of farmers	60	53	54	56	
	amount of manure collected (tons)	27	21	83	43	15,436
Cattle manure	number of villages	0	0	9	3	
	number of farmers	0	0	168	56	
	amount of manure collected (tons)	0	0	155.2	52	37,485

Experimenting with new options and sharing with others

At this stage of the process, farmers managed to acquire inputs to improve the fertility of their soil, some of which were new to them. Most of them did not want to apply the new inputs to all their land for fear of losing their production in case the untested inputs would damage their crops. For the research team, this provided an opportunity to initiate and discuss with farmers the concept of farmers' experiments.

Awareness on the concept of experimentation

Field meetings were organised with farmers interested in trying out new options to discuss possible experimental designs (see Figure 5). During these meetings, the research team and farmers would agree on a set of indicators that would be used to assess the performance of the different inputs on their crops in the fields.

These meetings were guided by the following questions:
- What do we want to learn from the experiments?
- How do we know if something has worked or not?
- What do we measure or observe and who does what?
- When do we take the measurements?
- What results do we share?
- When do we want to share this information?
- Who will be responsible for what?

During those meetings, the research team informed farmers that they wanted to monitor and assess certain parameters that included number of days crops take to emerge, tassel, silk and ripen; number of cobs; signs of nutrient deficiency; infestation by pests and yield (grain yield and mass of stover). Farmers, on the other hand, indicated that they would be interested in observing the *speed* (in terms of the number of days) at which crops responded to the different options and greenness or signs of nutrient deficiency. In this regard, it turned out that farmers were observing changes brought about by the different inputs in a broad context.

> 'I always thought that the issue of experimentation and designing of trials was a mammoth and complicated task and was the sole work of researchers. But having

Figure 5. A farmer demonstrating a trial on a small portion of land.

done this step practically with farmers, I can now understand the experimentation process much better than I would have learned this at a university or college' (An extension officer during monitoring of farmers' experiments, January 2000).

Although many farmers were initially enthusiastic about experimenting with several options, relatively few went beyond the planning stage and actually carried out tests in their fields (See Table 6). It soon turned out that some farmers preferred to see how trials could be designed and would later implement them independently in their fields. These farmers wanted to try out different options and combinations on their own. This is after they realised that trials made with the help of researchers were limited to certain options and conditions which did not necessarily correspond to the conditions on their own land.

During the monitoring of the experiments, there were instances where farmers shared what changes they observed in terms of soil structures and reduction of some soil pests where lime was used. This was a surprise to the research team because they could not detect those differences even though they were often visiting the fields. The researchers began to recognise that farmers had years of knowledge of their soils enabling them to see the most subtle differences in their land.

Table 6. Number and types of experiments tested by farmers (Ramaru et al.*, 2000).*

Type of experiment	GaThaba		GaMogano		Spitzkop		Total
	PF	CF	PF	CF	PF	CF	
Fertiliser[1]	-	5	4	-	4	-	13
Dolomite	2	1	-	1	3	1	8
Dolomite + fertilisers	1	4	3	1	6	1	16
Chicken manure	5	1	-	-	-	-	6
Chicken manure + dolomite	7	2	-	-	-	-	9
Kraal manure	4	2	-	-	-	-	6
Kraal manure + dolomite	4	1	-	-	-	-	5
Goat manure + fertiliser	-	-	1	-	-	-	1
Total	23	16	8	2	13	2	64

Note: PF= project field; CF= communal field.
[1] Fertiliser formula is N:P:K = 2:3:2 (N = nitrogen; P = phosphorus; K = potassium).

Farmers sharing with other farmers and outsiders

In addition to the usual field days that extension officers had organised for farmers in the past, the farmers - with the help of the research team - organised what is called mid-season evaluations (MSE). This event was a new concept for the farmers and researchers but they wanted to try it after they learned about it from farmers in Zimbabwe during an exposure visit. Hagmann *et al.* (1998) distinguish between mid-season evaluations and classical field days; in the former, all farmers within and outside the village are invited to see experiments in the fields whereas the latter is organised around a homestead or an office.

During the mid-season evaluation, farmers would share successes and challenges that they experienced in trying out different options to solve their technical problems in the field (Figure 6). For the first time, farmers shared with others how they managed to organise themselves into umbrella organisations to help them access inputs. These mid-season evaluations were so successful that they have turned into annual events organised by farmers.

Figure 6. Farmers from other villages were invited to share in the knowledge acquired during MSE.

Assessing the different options and planning for the coming season

The next step was for the research team to get farmers' judgement on the performance of the different options in the fields. Furthermore, it had to be decided how farmers would plan activities for the coming season.

Assessing technological options

The research team developed an assessment matrix on a large and laminated sheet allowing farmers to write technological options in a vertical column and criteria in a horizontal row. Depending on the number of criteria used to judge the options, a range of scores were developed to rank the performance of each option. Each group selected a spokesperson to present and explain the assessment of the options in a plenary meeting. The farmers would then analyse the differences and similarities across the different groups.

It was reported in the meeting that a number of farmers had carried out their own and independent experiments without the help of the researchers (see Table 7 and compare with Table 6). It was then decided that farmer-led experiments should also be included in the assessment exercise. It transpired that these kinds of informal experiments have always been done by a few farmers but there was no regular forum to share and discuss these innovations. In general, most of the farmers at the meetings developed more interest in technological options initiated by the fellow-farmer innovators and were willing to explore on them further the following season. These experimentation processes provided an opportunity for the farmers and the research team to engage in the learning process, something that had never been done before.

> 'I harvested 40 bags of sweet potato seed vines on the plot where I applied lime last season and harvested only 25 bags of vines on an area where lime was not applied' (Rosinah Libago, a farmers from Mbahela).

Table 7. Rating of soil fertility options independently initiated and implemented by farmers (Ramaru et al.*, 2000).*

GaThaba		GaMogano		Spitzkop	
Technology	Average score	Technology	Average score	Technology	Average score
Goat manure	12	Goat manure	15	Ash	15
Goat manure + dolomite	10	Chicken manure	15	Saw dust	14
		Malt + dolomite + fertiliser	13	Goat manure	13
		Cattle manure	6	Malt	11

Planning for the coming season

Village meetings to develop plans for the coming season started with recapitulating the results of the assessment of the different technological options. This exercise led to new discussions that either validated the assessment results or clarified what was discussed during the last meeting on the assessment of the technological options. The plans for the coming seasons were based on the following challenges:

- How best to spread the options that have proved to work?
- How can farmers strengthen their organisational capacity to access inputs for the options that are working?
- What is the best way to do more experiments on the options that are still questionable and who should do these experiments?

3.4 The spread of the soil fertility management innovation process

The sharing of knowledge during the organised mid – season evaluations has been the most effective means by which other farmers and officers have been exposed to the soil fertility management process. It is not unusual to hear farmers asking: *'When are researchers coming to do the same thing at our village?'* or *'When is our officer going to be trained in the Participatory Extension Approach and soil fertility management process?'*

The experience gained from the implementation of soil fertility management in the three pilot villages provided an opportunity for the research team to design a training workshop for other extension officers and scientists. This would enable other officers to get acquainted with the process and enable them to engage more farmers in their villages. Two workshops were designed:

- The first was a five-day workshop where the participants were taken through all the steps of soil fertility management. During the workshop, officers would be exposed to the laboratory setup to better understand the process of soil analysis. The officers also developed operational plans illustrating how they could implement soil fertility management in the villages that they supported.
- The same officer was given the opportunity to attend an advanced workshop in which a lecturer from the local university elaborated on the scientific aspects of soil science.

The farmers received a simplified version of the first soil fertility management training workshop that was given to the officers. They also managed to visit the soil science laboratory and had an opportunity to do some simple tests such as determining the acidity and alkalinity of the soil through a pH metre. Back in their villages, they shared the knowledge they acquired at a meeting organised by the community leaders. These farmers were then invited to:

- Train the local farmers groups in their villages and beyond

- Create awareness among extension officers and scientists who were not trained in the Participatory Extension Approach process. Realising that the farmers were able to be articulate about the process and were confidently sharing their experiences, the officers selected to be trained had a vision of what kind of farmers these participatory processes could produce.

From the assessment process conducted during 2004/2005, it was found that all the technical innovations had spread over all the districts of the Limpopo Province. For soil fertility management, Kganyago *et al.* (2005) showed that this innovation was implemented in 105 villages around the districts as follows: 24 in Capricorn, 74 in Vhembe, 2 in Mopani, 2 in Sekhukhune and 3 in Bohlabela.

3.5 Lessons from the reflection on the experiences

The reflection on the process of soil fertility management by the research team resulted in the identification of the success factors that were of importance for the unfolding process (Figure 7).

Subsequent reflection by the research team revealed that both the farmers and officers had gone through a process of joint learning during the development of the soil fertility management process. Table 8 documents changes that occurred among the farmers and officers at different stages of the soil fertility management process.

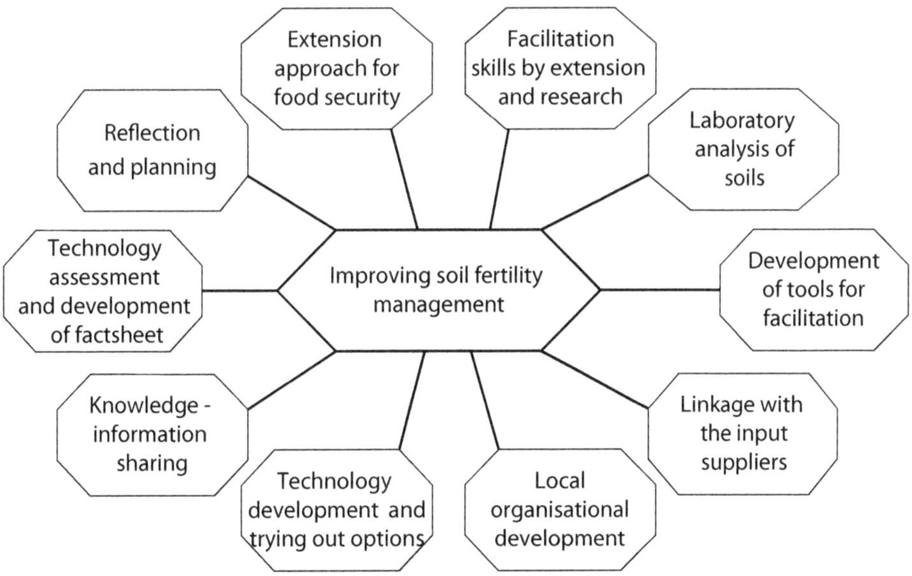

Figure 7. Strategy for soil fertility management in smallholder farming systems.

Table 8. The learning of researchers and farmers in the different steps of the process.

SFM process step	What changed and what was learned		How the process manifested itself
	Researchers	Farmers	
SFM step 0: PEA Training workshops of Researchers & Extension officers (EO)	Extension officers & scientist internalise some of the values such as recognising farmers / local knowledge, self-organisation, self-reliance	From participatory rural appraisal (PRA) workshops, farmers are aware of their situation and identify their technical problems	Mentoring and backstopping on the PEA provided by external consultants
SFM step 0: Soil Problem identification	Wanted to help farmers but did not know how to do it	Farmers were desperate and needed new types of fertilisers	Established research teams composed of extension officers and scientists to support new needs
SFM step 0: Planning the research process	After the PTD training, started to have a vision on how to start the process of SFM in selected villages		Plans were shared with the supervisors of the officers so that they would give logistical support
SFM step 1: Farmers' perceptions on soil fertility	Learned that farmers use local language to describe scientific things, e.g. yellowing of leaves	Farmers became curious to test their soils and volunteered to pay for the costs	Facilitation skills gained during the PEA helped officers to communicate scientific information to farmers through the use of tools
SFM Step 2: Planning soil sampling with farmers	Realised that farmers understand how to take soil samples with a spade rather than with an auger	Farmers were scared initially of experimenting but built up confidence as they practised	Farmers demand the results of soil analysis through feedback because they have paid for the costs of analysis
SFM step 3: Discussing the results of soil analysis	As researchers, we were happy that we managed to develop tools that represent scientific information in a simplified way	Farmers were happy about the results. The results correlate with experience from farmers	The extension officers who are involved are able to help other groups with interpretation of soil results

Table 8. Continued

SFM process step	What changed and what was learned		How the process manifested itself
	Researchers	Farmers	
SFM step 4: Linkages with service providers	We use to think that farmers didn't have money – because they were not buying fertilisers (mostly the communal farmers)	Farmers are existed to get discounts when buying inputs in bulks. In addition, inputs are delivered at not costs to the villages	Umbrella organisations have been established to negotiate with input suppliers for inputs on behalf of the farmers
SFM step 5: Designing experiments with farmers	'I can now understand the experimentation process much easier than I would have learned this at a university or college' (An extension officer during monitoring of farmers' experiments, January 2000)	More farmers come to learn how to design experiments but some are not interested in being involved in trials managed together with a research team	A core team of experimenters selected by farmers to try several options which will be shared during the mid-season evaluation
SFM step 6: Experimenting with new options and sharing with others	Learned that farmers see and measure parameters beyond the scope of our agreement (researchers & farmers)	Exposure to experimentation made other farmers initiate farmer-led trials	The mid-season evaluation caused more farmers from in this and other villages to get interested as well as try out options to solve their problems
SFM step 7: Assessing different options & planning new ones	Learned how to develop an assessment tool that helps farmers to assess their tested technologies	Farmers impressions *'It was the first time we tried some of these technologies and most of them did very well in terms of increasing the yield. In most of the cases, application of individual inputs did not give better yields as compared to when the inputs were combined (farmers from GaMogano, June 2002)*	Farmers are ready to choose those who will undergo training in soil fertility management – most of those chosen have participated in the trials

It is important to emphasise that the SFM process as a learning case was not dealt with in isolation. It was part of other innovation processes which were fostered within community development. For this purpose, PEA is seen as a guiding process and as a mechanism to help rural communities take charge of their own development by organising themselves, innovate and manage their own dynamic change, and bargain/negotiate with service providers (Hagmann *et al.*, 1998; Ramaru *et al.*, 2000).

3.6 Conclusions

The soil fertility management process is now implemented in areas such as:

a. high schools, as part of the demonstrations to help students understand the theoretical aspects of soil science in their syllabus; and
b. other donor-funded programmes (MacDev, LADEP & Land Care) to help their project beneficiaries to improve their soil fertility.

The facilitators who started with the process of soil fertility management are now facilitating other innovation processes (e.g. livestock mobilisation). They are functioning as learners in the new technical areas and as mentors in soil fertility management for new trainees. More farmers have been trained by the soil fertility management facilitators with the help of other farmer trainers. The farmer-trainers are serving as coaches and mentors of newly trained farmers during the facilitation of soil fertility management in the communities.

Also relevant to other innovation processes (soil and water conservation, small-scale seed production and livestock management) which were developed during the implementation of the Participatory Extension Approach process, was the fact that the innovation process unfolded as the result of continuous joint learning among the stakeholders involved. Success factors for the implementation of these innovations within the framework of Participatory Extension Approach are described as follows:

- Facilitation skills are import to start a process of innovation in the community. Where possible, these skills can be learned not only by the extension officers but also by the scientist.
- Within the research team joint planning is encouraged in order to enhance learning from peers in the group. The importance of preparing and sometimes rehearsing each step together within a team of facilitators cannot be overemphasised.
- A reflection meeting by the research teams after the village workshop is also important so that those who acted as facilitators could improve on the weaknesses that might have been identified. Continuous sharing and giving feedback should be the group norm to facilitate learning from each other and to enhance conceptualisation of the lessons.
- Awareness creation at all levels of the innovation process is important for farmers to be able to take informed decisions on what to do next. This can best be done with tools designed to simplify complicated scientific information.

- Also important are the exposure visits by the community representatives to sources of innovations within and outside their localities, as well as giving feedback about the findings to the community at large. This exposure is important to enable the research teams to change their way of doing things.
- Strengthening the local organisations is important for linkages with external service providers and input suppliers. They also play an important role in the monitoring of experiments and the organisation of sharing events.
- Experimentation is a tool for joint learning and can be fuelled by awarding prizes for best experiments and selection of promising technologies for further trials in the coming seasons.
- The pillar of any innovation rests in the available mentoring/coaching mechanisms. Where this does not exist (yet), an attempt should be made to develop it at the level of researchers and farmers.

It is important that facilitators of this process open their ears and document any striking feelings, concerns, appreciations and comments from farmers about the process and developments resulting from their involvement.

References

Allen, W., M. Kilvington and C. Horn, 2002. Using participatory and learning - based approaches for environmental management to help achieve constructive behaviour change. Landcare research contract report: LC0102/057. Report prepared for the Ministry of Environment, Wellington.

Anonymous, 1995. White Paper on Agriculture. Northern Province Department of Agriculture, Pietersburg.

Borman, J., J. Ranwell, G. Venter and L. Vosloo, 1989. Fertilizer handbook, Fertilizer Society of South Africa. Hennosmeer.

Hagmann, J., E. Chuma, K. Murwira and M. Connolly, 1998. Learning together through participatory extension: A guide to an approach developed in Zimbabwe. Published by the Department of Agriculture, Technical and Extended Services, GTZ and ITDG Harare, Zimbabwe. 59 pp. Available at: http://www.gtz.de/agriservice/areas/concepts/concepts.html#7 or modified under http://www.odi.org.uk/agren/papers/agrenpaper_94.pdf.

Hagmann,J., E. Moyo, E Chuma, K. Murwira, J. Ramaru and P. Ficarelli, 2003. Labout developing competences to facilitate rural extension processes. In: C. Wettasinha, L. van Veldhuizen, and A. Waters-Bayer (eds.) Advancing participatory technology development: Case studies on integration into agricultural research, extension and education. IIRR/ETC/CTA., pp. 21-38.

Kganyago, E., T. Ngomane, J. Nkum, G. Ramm, G. Durr, R. Kressirer and N. Schaffer, 2005. Project progress review (phase 01/2003-12/2006). Focal Area Local Governance and Development, BASED/DRS. June 2005.

Marais, D., J. Taylor and A. Kaplan, 1997. Action - learning for development. Use your experience to improve your effectiveness. Juta and Community Development Resource Association, Cape Town.

Moyo, E. and J. Hagmann, 2000. Facilitating competency development to put learning process approaches into practice in rural extension. In: FAO (ed.) Human resources in agricultural and rural development 2000, FAO, Rome, pp. 143-157. Available at: http://www.fao.org/DOCREP/003/X7925M/X7925M14.htm.

Ramaru, J., Z. Mamabolo and J. Lekgoro, 2000. Improving soil fertility management in South Africa: Learning through participatory extension approaches. Managing Africa's Soils No 19. IIED, London. Available at: http://www.iied.org/drylands.

Ramaru, J., J. Hagmann, E. Chuma, P. Ficarelli, M. Netshivhodza and Z. Mamabolo, 2004. Building linkages and bargaining power between smallholder farmers and service providers: Learning from a case on soil fertility inputs in South Africa. Ugandan Journal of Agricultural Sciences 9: 204-214.

Schuh, C., 1999. Analysis of farming systems in selected areas of the northern province (South Africa). Volume 1, Main Report, BASED, Pietersburg.

Shao, X., V. Konovalchuk, B. Clark and T. Bruening, 2004. Identifying problems facing smallholder South African farmers through participatory rural appraisals - Case studies with smallholder farmers. Proceeding of the 20th Annual Conference of AIAEE, Dublin, pp. 588-597.

Van Veldhuizen, L., A. Water-Bayer and H. De Zeeuw, 1997. Developing technologies with farmers: a trainer's guide for participatory learning. Zed Books, London.

4. From top-down to reciprocal services[6]

A case of exemplarion action research

Leni Beukema and Ben Valkenburg

4.1 Introduction

In this contribution we present a research project on processes of transformation in organisations for elderly care in the Netherlands. In the first section we present the central issue of the project, a demand-driven approach to elderly care. Our first research activities on this issue were based on more or less regular research methods. However, it soon became clear that further research would require action research. In the second section we summarise the central arguments for this: how, when and why did our research become action research. In the third section we present the method of exemplarian action research we used.

The relationship between researcher and co-researcher is essential in a project like this one. On a theoretical level the members of the researched party are seen as co-researchers. On a more practical level the questions as to who can be regarded as co-researchers, what role they play in the research project and how they can fulfil this role, have to be answered time and again, with different answers in different situations.

In the remaining sections of this article we describe how we attempted to deal with these topics in our everyday research activities.

4.2 A demand-driven approach as empowerment of both clients and workers

In the Netherlands the welfare state has a long-standing, deeply rooted supply-driven tradition. Policy makers (and professionals) define the problems, formulate the solutions and shape the provisions. Often the basis for doing this is knowledge that is seen as more or less objective (compared with the knowledge of citizens) and general (compared with the specific context of citizens). Only in the process of policy delivery does the individual client enter the picture. In this stage of policy delivery, the role of individual clients is nevertheless limited. They are supposed to co-operate in a process that is not primarily based on their own definitions of problems, analyses and strategies, but on those formulated by the political process at a central level. There is usually little room to actively discuss these definitions, analyses and strategies.

[6] This contribution is based on an extended report on the research project in the Journal of Action Research, 2007, Volume 5(2): 161-180.

What applies to the welfare state in general, also applies to elderly care. In this supply-driven tradition the power of institutions over the lives of clients offers the latter very little room for manoeuvre. Rules are set for the timetable of the day and professionals arrange the treatment of 'their' clients. First and foremost, a demand-driven approach is developed and discussed as an alternative for this supply-driven approach.

A demand-driven approach can be viewed as a possible answer to the growing reflexivity of individual and social life (Giddens, 1990, 1991; Beck and Beck-Gernsheim, 2002; Valkenburg and Lind, 2002). What do we mean by reflexivity? In modern society, knowledge is permanently used, discussed and developed in everyday practice; it involves a process in which the meaning of knowledge is continually changed within the context of everyday interaction, in the mutual exchanges between people. If the contexts are different, this reciprocal process may also be different. This changing meaning of knowledge is an important factor in processes of individualisation. At the same time it forms the outcome of processes of individualisation. Based on a permanent flow of new knowledge and in different contexts, people develop their individual identity and actions. In leading their everyday life they further develop their knowledge. In this reflexive process what is 'true' for their everyday life today, may become 'untrue' tomorrow. Individual identity, and the everyday life people live, has become a reflexive project. In this discourse a demand-driven approach primarily means that social policy, including elderly care, should link up with and do justice to the reflexive projects of individual people in a reciprocal way. Put differently: the provisions and care should make sense and be of use within the actual interactions and exchange processes people engage in their daily lives. The starting point for doing this is where policy and the everyday life of individuals intersect, i.e. in the dialogue between the professional that delivers policies and the individual citizen.

In the elderly care sector the urge to implement a demand-driven approach is felt to be very high, not only because of the social tendencies mentioned above, but also due to changing views about age and aging. In the West age isn't seen as a source of richness and wisdom, but rather as the road to decay and dependency (Niessen, 1995; Warners, 1998, among others). In elderly care this is one of the main pillars of the deficit model, in which the dependency of old people is stressed and sometimes even increased by the institution. Opposed to this are views of aging in which individuals are primarily defined as actors that may have problems, but who are, nevertheless, active and competent. The modern motto would seem to be: 'forever young'. We now speak of the fourth and fifth phases of life. If we are to believe the advertising campaigns of some insurance companies, then the fourth phase – between 55 and 75-80 – would seem to be the best phase of all, and it is in the fifth phase that 'real aging' comes to the fore. From this perspective the central notions for elderly care are health, independence and the ability to run one's own life.

Since the eighties, demand-driven care has therefore become a major phenomenon in policy in the health care sector in the Netherlands. However, this approach implies a transformation with far-reaching consequences. In the primary process - the everyday interactions between professionals and clients- the professional has to link the complex and differentiated situation, reflexivity, needs and competences of the individual client on the one hand, and the differentiated forms and instruments of care on the other. In a demand-driven approach the professional can only to a very limited degree base this primary process on general ideas about clients or on more traditional forms of professional knowledge. This way of dealing with clients is part of a supply-driven approach, in which general ideas about clients are translated into uniform provisions that are then delivered to individual clients. In a demand-driven approach, the professional has, time and again, to investigate complexity and differentiation on the level of the individual client. This is possible only if the professional is able to develop a dialogical relationship with the client. Furthermore, a demand-driven approach in the primary process not only has to be supported and facilitated by the secondary process (management, policy development in the organisation, the organisation of care), but also has consequences for this secondary process as such. A demand-driven approach requires a dialogical relationship between the primary and secondary processes, and thus reciprocal forms of management. Therefore, a demand-driven approach requires a long-term process of implementation, in order to empower clients, professionals and management to adapt to new requirements.

In the second part of this article we deal with these consequences more extensively.

4.3 Preliminary research

As we said in the introduction, more or less regular research methods were used in our first research activities into a demand-driven approach.

In 1998 we were asked by RAP-Rijnstreek[7] to conduct research into the consequences of demand-driven care for the working conditions of workers and personnel management (Human Resource Management) in the elderly care sector. At this stage the question was a descriptive one: what is happening and what are the different perspectives on the subject. We carried out a telephone survey among management and workers' representatives in fifty homes for the elderly. Furthermore, we carried out a number of case studies, in which we looked more closely at four organisations. The conclusions were that although demand-driven care was prominent in visions and mission statements of the organisations, it had barely been implemented at all in the primary processes of the organisations. Workers had not been trained to work in this way. As a consequence, a task-driven organisation of work – based on the institutionally determined supply of services - continued to be the dominant principle when organising care. Human Resource Management was underdeveloped; only 25% of the organisations had some sort of HRM officer.

[7] RAP-Rijnstreek is an organisation that focuses on labour market developments in the care sector and aims at improving the link between demand and supply in this labour market and strengthening HRM in the sector.

The results of the preliminary research and case studies led to a discussion on how to improve the situation. Our argument was that demand-driven care should first and foremost be implemented in the everyday practice of workers in the primary process. Therefore, we made a plea for creating an articulated platform around the primary process and giving it a central position in the transformation process. Furthermore, our argument was that demand-driven care presupposes a demand-driven approach on the part of workers too. Our advice was to arrange for a field experiment, in which new job specifications for workers could be formulated, and adequate competences and planning methods to facilitate the implementation of demand-driven care could be developed.

The board of RAP Rijnstreek followed our advice and started looking for elderly care homes that could be involved in this kind of experiment. This was not without problems. In order to be involved in such an experiment, demand-driven care had to be put high on the agenda of the organisation. Management had to commit itself to investing time and money in experiments with a demand-driven elderly care, in a situation where both were scarce. There were not many managers who were willing to do so.

We finally found an organisation. Recently three different homes for the elderly had merged into one organisation. They were looking for ways to handle the different cultures of the homes involved in the merger and saw the proposed experiment as a fruitful way to approach this. In this stage of setting up an experiment, the relationships between researchers and management began to shift from a researcher–customer to a researcher–co-researcher relationship.

At first management saw the researchers as experts who could tell them how best to handle their situation. It gradually became clear that this was not adequate. This was partly the result of our interventions as researchers. Setting up an experiment and research project requires a dialogue about the assumptions they will be based on. In this dialogue we as researchers clarified our own assumptions, tried to make it clear that we were not experts in the above sense and proposed a different approach to the research project. To make this more than an academic debate, the researchers' arguments have to be linked with the everyday experiences of, in this case, the managers concerned. In the dialogue and the first stages of the research project this connection was established along two lines.

Firstly, it gradually became clear that the evaluation and transformation of everyday practice required a clear and shared vision of a demand-driven approach. Like many other managers, our managers had had negative experiences with 'visions'. They often 'look nice, but have no effect at all'. The discussions between researchers and management made it clear that this is not primarily caused by unwillingness on the part of workers to conform to a new vision, but rather by the way these visions are formulated. They are often formulated at the level of the general policies of the organisation and present general assumptions that everybody can easily agree to. Once they are formulated, the 'rest' is seen as a matter of implementation.

This approach fails to do justice to the fact that workers too have visions, and that these visions are at the basis of their everyday working processes.

Once they started to look at their own organisation from this perspective and with the support of our research activities, our managers realised that 'under the surface' of shared assumptions, very different visions were practised. The conclusion was that the development of a general and shared vision required that general assumptions be investigated in terms of their practical implications in the specific context of the organisation(s) and vice versa. Furthermore, researchers and co-researchers agreed that the visions that are held on the different levels of the organisation should be related to each other in a reciprocal way.

Secondly, it became clear that the possibilities for implementing a demand-driven approach should also be developed in a reciprocal way. This process concerned not only the workers in their direct interaction with clients; it also called the routines of management itself into question. It stressed the inadequacy of top-down management and of applying very strict procedures and protocols on workers. Management was challenged to create conditions for workers to take action on their own initiative and to take responsibility for uncertain situations. Moreover, it became necessary to stimulate workers to be more reflexive about their own actions, their underlying motives and about the (intended and unintended) consequences of what they do.

These changing demands on management had to be met in different situations. Management in the three homes had, up until then, functioned quite differently, and varied from mainly top-down management to participative management to hardly any management at all. Consequently, routines in the homes and starting points for discussion on possible changes differed widely.

In terms of content, the results of the experimental study were a set of integrated theoretical concepts on the demand-driven organisation of elderly care and more methodical instruments on the implementation of these concepts in different contexts. The vision on demand-driven care was better clarified as well as its relationship with the position of workers and with the organisational structure and culture. A dynamic planning cycle to help implement demand-driven organisations in the care sector was one of the key outcomes of this research.

In terms of the research process, it gradually became evident that an approach in which the researchers develop knowledge and give advice that can then be implemented, was inadequate. The implementation of a demand-driven approach required an approach in which the knowledge of the researchers and the researched party are combined in a reciprocal way, and in which the implementation of change was used as a research instrument. This is only possible if managers and workers (and clients - we will return to this in the second part of the article) have a role as co-researchers, who investigate and discuss their own visions, the ways they put them into practice, and the opportunities for change in this everyday practice.

For the researchers this means that they are able to move freely in the three homes and start up discussions with all those involved (operational management, workers and clients) on the topics under study. They have the opportunity to build a situation of trust with workers (and clients), in which the latter are willing and able to talk openly about that everyday practice and the visions, motives and ideas underlying it.

In discussions between management and researchers on this and similar points, the reflexivity of management increased and they grew in their role as co-researchers. For them this led to the conclusion that they could handle the further implementation of a demand-driven approach themselves, together with workers and clients. We as researchers had to find another setting for further research. Based on the results to date, this further research was based, from the outset, on the methods of (exemplarian) action research.

4.4 Exemplarian action research: implementation of demand -driven care

The results of the preliminary research raised interest among managers in other homes for the elderly in implementing demand-driven care in their own organisation. The researchers were asked to co-operate in this implementation. In 2002 we started to focus on one of these homes, which was part of a larger organisation spread over twelve locations. In due course word spread and while writing this contribution we have extended our research to cover four other locations belonging to the same organisation.

From the start, the research project was based on the principles of exemplarian action research ('exemplarisch handelingsonderzoek'), a version of action research that is very much inspired by the works of Freire, Negt and Giddens (Coenen, 1989, 1998; Beukema, 1996, and others).

In exemplary action research we distinguish three stages: the thematic, the crystallisation and the exemplary stages (Coenen, 1996: 14/15; Figure 1).

In the thematic stage the primary aim is to get to know the situation under study. The main source for this is the knowledge of the actors involved (interviews, discussion). Furthermore, the researcher can use written material, observe the situation, attend meetings, etc. As a result of this stage the researchers will not just describe the situation. They will also try to structure it: what are the central issues, which problems should be solved, what opportunities are there for action and which competences are available or can be developed. In other words: the researcher's aim is to structure the reality under study into themes, not only by describing this reality but also by analysing it in terms of opportunities for change.

The crystallisation stage consists of two parts. In the first part the researcher and co-researchers discuss the accuracy and the adequacy of the analysis made in the thematic stage. By discussing the importance and possibilities of the various themes, they try to establish

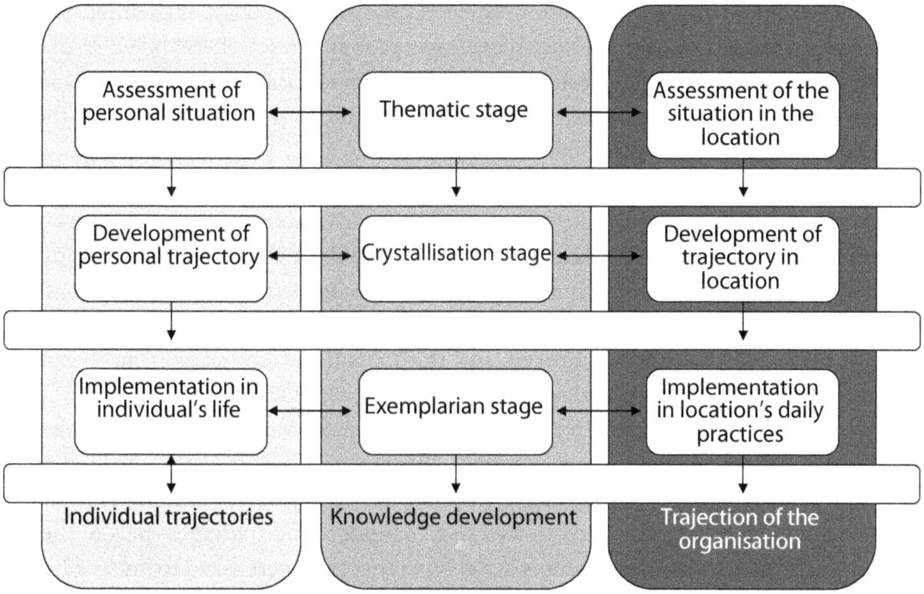

Figure 1. Research strategy. Formulating shared vision on demand-driven care and its implementation process.

the most important theme, the *exemplar*. If researchers and all co-researchers involved agree, the second part of this stage starts: the (co-)researchers draw up a plan to deal with the exemplar through consultation.

The exemplary stage begins when this plan has been sufficiently well developed for implementation. In this stage the implementation process is at the same time a goal in itself and an instrument for further research: what intended and unintended consequences occur during the implementation of the plan, how do people react to these consequences and what further strategies can be developed to enhance the process of change and reduce obstacles to this change.

These three stages can be repeated several times. When conducting this kind of research in the practical context of existing organisations, these stages are not always separated as rigidly as it appears on paper. Organisations consist of many actors that do not always follow the same route at the same time. Moreover, unanticipated events may require a shift in stages.

4.5 Relationship between researchers and co-researchers

The team of researchers on this project consisted of a project leader and eight researchers –different combinations at different locations. This team cooperated with co-researchers from different levels in the organisation. In every location the stages of exemplary action research were contextualised and within these stages groups of researchers and co-researchers could be different. In some locations emphasis was put on cooperation with operational management and client representatives, and in other locations workers also took on the role of co-researchers. However, some main lines can be drawn.

In each of the five locations the initial contact was with management. They were asked to investigate the situation in their location and, based on that, they were responsible for the implementation process as a whole, for the finance, and for the accounts to the firm's strategic management. Subsequently, a committee was set up at every location that was responsible for the research process and the steps that were taken. The researchers stressed the importance of a broad representation in these groups (management, workers, and clients). This was achieved in three locations, and in two locations the committee consisted only of management representatives. Workers and clients were consulted on the research process separately by the researchers. In every location representatives of clients, management and workers were interviewed to give input to the analyses as part of the first, thematic stage. They were co-researchers in the sense that they were asked to investigate their situation, actions, motives, visions and ideas about change in an open interview situation, together with the researchers.

In each location workers (depending on the size of the location - all workers or those workers who were considered capable of disseminating their insights among colleagues) were trained to reflect on their own practice, on the way they attempted to deliver demand-driven care and on the opportunities for improving it. This training was organised in several sessions over a period of one year. In this way, the role of these workers as co-researchers was further developed. They became part of the research activities on a more permanent basis.

The researchers had multiple roles in the process. Apart form the traditional roles such as interviewing and writing reports, data gathering and analysis, they also performed the role of discussant, trainer, coach and consultant. To raise the quality and validity of data, a central focus in the actions of the researchers was the building up of trust with all the different co-researchers. Furthermore, the participants involved do not automatically take their own knowledge and actions as being valid and essential to the research and change process. They do not spontaneously see themselves as 'co-researchers'. The researchers have to convince them to *become* co-researchers, to trust their own knowledge as much as that of others, and to dare to act, with all the intended *and* unintended consequences that it entails, and to make errors while doing so (just as the researchers themselves do). Consequently, this implies that the researchers have to take on an open attitude and dare to be vulnerable in the process and explicitly reflect on their own actions as well as on those of the co-researchers.

In our research we had to deal with workers, managers and clients in elderly care, with all the power relations that this involves. In all the cases, management was and remained the first and most important contact. However, management also holds power and when relations between workers and management are not too good, the relationship of the researchers with the managers might have a negative influence on the relationship of the researchers with the workers. As far as we could ascertain, this situation did not arise very often, although in three larger locations management functioned at quite a 'distance' from the workers. We tried to cope with these differences in power relations by explicating our assumptions at the start of the research, but mainly by developing long-term relationships of trust with different key actors at different levels of the organisations under study. This sometimes meant the research had to change focus to improve relationships between workers and management, because the workers asked us to do so, as we set out in one of our examples below. In the situations where relations weren't too good, 'room to manoeuvre', as we called it, is essential: management has to give the researchers the space to disagree and have their own responsibility in the change process. So trust between management and researchers is a precondition in this kind of research in order to build up relationships of trust with other members of the organisation. In this sense the researchers also have the responsibility of building such a relationship of trust with operational management and workers concerned.

This relationship of trust is itself in question because of the differences in power between the researchers and the workers involved, as Lennie *et al.* (2003) have convincingly shown. In our research, our aim was not for equality but for equivalence: both parties have their own position in the process of action, learning and building knowledge and they value and respect each other in these positions. In the following sections we return to the relationship between researchers and co-researchers in the different stages of the research project.

4.6 Formulating a shared vision

To be able to formulate a direction for change, a discussion about the vision with strategic and operational management at location level is necessary. McNiff, after Sowell, speaks of the attractiveness of unconstrained vision, which sees beyond the stable state and imagines ways in which it might be improved (McNiff and Whitehead, 2000: 11). In our opinion this vision doesn't imply a new 'model' or a desirable blueprint for the future. A shared vision is also needed to shed new light on the existing situation, by defining new meanings, power relations and values in a future-oriented, but nevertheless realistic way. In this sense it is a condition for quality and validity.

In order to do this as explicitly as possible, the researchers have two issues to deal with.

First there is the issue of scientific knowledge. In our case we as researchers had previously explicated our vision on demand-driven care in several publications. It takes time and effort to make it clear that this explication is not a pre-set prescription for change, but a guideline for discussion. It has to be adequate, that is to say contextualised, actualised and made to fit the local situation. This requires reciprocal adequacy: the input of the everyday knowledge of the research participants and their willingness to act as co-researchers. So every presentation of the research in the different locations was combined with discussion and training sessions on the usefulness of such a project in the specific context. Discussion soon became very lively, especially in those cases where we were able to present a report on demand-driven care in the location as a result of preliminary interviews. The role of the researcher in these discussions was focused on asking questions to clarify certain points, on structuring the different opinions that were put forward in the discussion and on opening up thoughts about opportunities for change. This way, the researchers try to break through the dominant view on 'expert-knowledge' as general, value free and superior knowledge that can be implemented from the top down. Some managers are in a sense waiting for this breakthrough and sympathise with the researchers' approach. Others expect something else:

> 'Please tell us how we can handle this demand-driven care. What are best practices elsewhere with your approach and can we get an instrument to implement this at our location?'

If they do not have this idea about scientific knowledge, they often go to the other extreme in which scientific knowledge is merely instrumental:

> 'I want to take the following steps and I need you to fill in our training. We'll take care of the rest.'

These two reactions resemble the observation made by Fricke (in: Brydon-Miller *et al.*, 2003: 19) who warns us to avoid difficulties when bridging the two worlds of theory and praxis: the researcher will be reduced to either an academic or a consultant. These routines of management, however, are quite understandable. In their experience academic knowledge implies analytical reports, full of 'superior' knowledge that barely gives them a clue about how to handle the problems they are faced with in their daily practice. If the researcher comes to support change, he comes as a consultant. In this situation reciprocity in the relationship between researcher and co-researchers has to be developed, which takes time and effort. It implies discussion about existing routines and investigation of their, intended and unintended, consequences for change. A precondition is that researcher and co-researchers trust each other. This too is something that has to be developed over time and it is not always successful. Some managers may prefer co-operation with other researchers or the researcher may see insufficient points of departure for a reciprocal research project. In this project this situation actually arose in one location: management and researcher were unable to find common ground for cooperation and the location did not participate in the project.

Secondly, the discussion is an attempt to get an adequate view of the consequences of the vision in the specific context. As we clarified before, making the shift to demand-driven care requires new actions of workers in the primary process and in the way this process is managed and coached. But the attitudes, competences and actions of management and the willingness to reflect on them with an open mind, are also at stake (see also McNiff and Whitehead, 2000; Argyris 1990, 1999).

> 'When we started this project I thought I knew what it meant that I also needed to change. But at that time I didn't realise how far-reaching the consequences would be' (manager).

As a result of these discussions between management, researchers and in some locations also workers and client representatives, the central questions for further research were formulated:

- What are the main principles of demand-driven care, both on the level of the primary process, and of organising and managing the organisation?
- How do clients, workers and executive and strategic managers see the current situation in the primary process? What are the strengths for transforming it into a demand-driven approach and what problems have to be solved?
- How do clients, workers and (strategic and operational) managers see the possibilities to design demand-driven work and which problems have to be solved?
- What are possible starting points for the transformation towards demand-driven care and how can these starting points be transformed into activities?
- What starting points does a demand-driven approach of clients and workers offer for the set-up and planning of the organisation?
- What lessons can the organisations learn from each other, and what lessons can be learned for the organisation as a whole?

4.7 Looking for relevant themes (thematic stage)

The research goal of this stage is the inventory of themes that represent a shared understanding by researchers and co-researchers of the situation under study. It not only concerns a shared problem definition, but also the normative context and the different perspectives in power relations. The researchers get a first insight into dominant routines in the primary process (interaction between client and workers, and between workers themselves) and in the secondary process (interaction between workers and operational and strategic management and the organisation of work). This insight is a necessary condition for any programme of change. If this connection is missing, individual members and groups within the organisation will not be able to be part of the implementation process, or may deliberately withdraw from it. In both cases this may put an end to the process as a whole. To develop this insight the researcher must have room to manoeuvre in order to build up trust with different kinds of people who, from different perspectives, make up the organisation. The quality and validity of the research process is therefore a precondition for organisational learning and adequacy in implementing demand-driven care.

This is not always easy, especially when there is tension between management and workers, either individuals or groups. The researchers do not yet know the underlying processes and relationships within the organisation and therefore have to be careful when proposing or undertaking actions. As you might imagine, in this stage the attitude of management towards other members of the organisation and towards the researcher is very important.

This stage shares essential features with 'traditional' qualitative research, where the main focus is on understanding and adequately analysing the situation. Our research methods include interviews, participatory observation and discussion. The interviews with clients, workers and management were primarily used to acquire an initial view of the different perspectives of the members of the organisation. The respondents were chosen carefully. They have to cover the different hierarchal levels and the different departments. They also have to represent the different opinions and discourses within the location. In larger locations we had to take a sample based on these criteria. In smaller ones we conducted in-depth interviews with some workers, but had talks with all. In this stage participatory observation is an important method of acquiring information about everyday interactions, about the routines underlying these interactions and about their intended and unintended consequences. Discussion is used as a research method where researchers and co-researchers can learn from each others' competencies: co-researchers give their views on their everyday practice, researchers present their views by explicating the different lines in the discussion and their link with demand-driven care. Moreover, discussion also gave an insight into the relationships between different members of the organisation and into their shared vision on the situation within the location. These discussions can be plenary between all members of the organisation. In this stage, though, it is better to form groups where people feel safe to speak freely. When, for instance, we had a kick-off conference with all the members of the organisation, we also created the opportunity to discuss things in small, self-selected groups. The primary aim is to create a situation in which people are willing and able to speak freely, and therefore realistically, about their actions, thoughts and ideas.

The result of this stage is a written report, with a thematic presentation of the point of departure for the further implementation of demand-driven care. Clients, workers and management as co-researchers are asked whether or not the researchers' interpretation is adequate and complete, in order to validate the research results and improve their quality. This 'check' can be risky. Often the researcher is supposed to 'know best'. Furthermore, many people at this stage want to be polite so do not question the work of the researchers. All this is understandable, so it is primarily the task of the researchers to create opportunities for discussion and criticism and for expressing doubts about their interpretation. This can be done in different ways, for instance by putting different views on the same problem on the table, by asking further questions, and by creating an atmosphere of security and trust for each individual.

In one of the locations, for example, we presented our findings on the relevant themes within the location. Everybody agreed. When we asked for the most important theme, almost everyone came up with a different answer. Underlying processes within the organisation turned out to be highly influential on the vision of demand-driven care and the related themes. So, lack of trust on the one hand and almost no shared responsibility for a client's well being on the other, had to be tackled in the process of implementing demand-driven care. The discussion about the initial research results explicitly brought these processes to the table. A new round of discussion was organised, first in multidisciplinary groups, then in mono-disciplinary groups, in order to obtain a firmer basis for the intended change.

4.8 Discussion on the themes and making a plan of action (crystallisation stage)

Discussing the possibilities for change and priorities sheds new light on practices, routines and problem definitions, and in this way on barriers and possibilities for change, both on individual and organisational levels. In this stage individual coaching and group discussions are important for increasing reflexivity on both levels in an atmosphere of trust, as an important condition for implementing demand-driven care.

One of the important issues in this stage turned out to be 'professionalism'. Supply-driven care corresponds with a traditional view of professionals as experts. They know what is best for the client. Their interaction with the client is based primarily on this professional knowledge. In practice this implies that the professional also has the 'power' to impose his ideas on the client. Many professionals fear that demand-driven care will change this relationship into the other extreme: the client tells the professional what to do. This would not only have an effect on the status of professional knowledge, but also on the position of the professional in his relationship with the client, on the way power is handled and on the 'professional attitude' of keeping a distance from the subjective perspective of the client. It requires considerable discussion to explain that demand-driven care does not reduce professionalism as such, but places different demands on it. It requires a dialogue between client and professional, in which asking adequate questions, the input of the competencies (and knowledge) of both client and professional, and shared reflection are all linked to support the development of the client. The discussion on these key competences to do this opens up possibilities to formulate a new kind of professionalism, where relational and reflexive competences are just as important as the professional knowledge about care. This discussion should be on group level (general ideas, perspectives and organisational possibilities; the right-hand side of Figure 1) and on an individual level (implications for individual professionals and their coaching; the left-hand side of the figure). The combination of group discussion and individual coaching is a very powerful one for increasing possibilities of change with substantial support from all members of the organisation. It also contributes to the quality and validity of results, in the sense that it permanently connects, or reconnects, general discussions to the specific situation of the people concerned.

Action plans are designed in different ways, depending mainly on the existing and desirable culture of the organisation. In most cases this was done in working groups, consisting of co-researchers suitable for the task at hand because of their role in the organisation. Reformulating the primary process or designing new function profiles, for example, requires the involvement of different people as management, care providers, Human Resource Managers, etc. Designing new formats for dossiers requires groups of people who work with these dossiers, and preferably include the clients themselves. In this process, the answer to the question about which members of the organisation are seen as co-researchers and what their role should be, becomes specific and differentiated, depending on the issues at stake. In other cases training is also incorporated in the process. Furthermore, forms of intervision have been used, in which discussion about demand-driven care goes hand in hand with small changes and reflection on their effects. In these cases the researchers bring in their general knowledge of demand-driven care and organisational change, and by relating this to the everyday practice of the co-researchers facilitate the learning process. In this situation the researchers are aboth trainers and researchers, because these sessions provide a lot of interesting new data.

During some training sessions, for example, workers and operational management discussed differentiation a lot. They work with different clients for whom different questions should be asked. This interaction takes place in different situations: people's apartments, an intramural setting or a closed setting. The researcher attempted to structure all this information by formulating different roles of professionals in different situations and brought this back in the training group. When asked whether this was an adequate structuring of the former discussions people were initially surprised that they had developed these ideas themselves. Then a real discussion started on the adequacy of the work of the researcher. To conclude, the researcher adjusted the findings and visualised them, after which the new knowledge could be disseminated among colleagues and management.

In the process of designing an action plan the researchers have a threefold task. They have to facilitate the design process (see above). At the same time they have to analyse this process in terms of the underlying developments in the organisation that need to be handled, such as the theme of trust and responsibility as mentioned before. Here the process of implementation explicitly requires a discussion with management and steering committee as co-researchers. Thirdly, their output is substantial in terms of new products of demand-driven care that have been developed during the research process. In our experience a considerable amount of creativity emerges when co-researchers feel free to experiment with a new way of working and we see it as a task of the researchers to systematise these activities and formulate them in terms of products. Demand-driven care here gives a specific interpretation of entrepreneurship in the care sector, where products are developed by workers and clients together. New methods of food distribution, training programmes for volunteers, the setting up of a bar-cum-theatre for residents and elderly people from the neighbourhood are just some examples of these innovations.

During this stage the role of the researchers changes from the more traditional role of data gathering and reporting towards a more active role as a facilitator of the research process by working with the co-researchers on plans for redefining the primary and secondary processes of the organisation. For instance, knowledge of intended and unintended consequences of change gathered from training sessions is first taken to management level by the researchers. Later on researchers and workers and management as co-researchers need to renew communication lines within the organisation in order to strengthen the possibilities of reciprocal communication within the organisation.

Most of the time the crystallisation stage is not a linear process. It has to link up with, and in this sense has to follow, the learning processes of the different actors involved. At the same time, when individual actors withdraw from the collective process they can inhibit it, depending on their position in the organisation.

In one case, management changed just after the thematic stage had ended. This meant that the research activity slowed down somewhat because it had to be re-connected to the new situation and the new managers involved. In another case, discussion about demand-driven care was questioned by workers who were most interested in drawing attention to the top-down communication within their department. They first needed to establish trust in order to solve this problem, before discussing and changing their own professional attitude. A small thematic stage on the level of the department was needed in order to gain new insight into the relationships in this department and to start discussion with workers and management about solving the problems at hand.

4.9 Exemplary stage

The crystallisation phase ends when an exemplar is formulated. An exemplar is defined as the most important theme, according to researchers and co-researchers, that should be tackled first and foremost. There are three criteria. It should be a central theme within the organisation; to handle it is a condition for handling other themes. It should also be realistic, in the sense that it must not only deal with problems, but also with strengths and realistic aims in the given context. Last but not least, it must be relevant for the learning process of the organisation; what is learned in one situation must have exemplary value for others.

Processes of intervention and evaluation are central to this stage of action research. The research goal is to implement the plan of action and the evaluation of action and learning in order to secure the change process. Moreover, the aim is to enable the members of the organisation to reflect on their own situation and the presumptions of their own individual and collective actions. In other words, research in this stage is explicitly connected to the learning organisation, where co-researchers are able to see their activities and the things they learn as a research process. This is also the stage where researchers prepare for their withdrawal and teach the co-researchers to deal with their situation themselves as a continuous research process. Once again: guarding quality and validity is not only an explicit task of the researchers but becomes more of a shared task.

The research methods again are participatory observation, intervision and discussion, topped up with different evaluation methods such as a written evaluation by workers of the courses they participated in or discussion about a qualitative evaluation report within the steering group.

As we write this contribution we are working on exemplars in two locations. In one of these cases management has chosen to take the implementation into its own hands. We intend to perform the evaluation of the process after one year of experience.

The first exemplar was the development and implementation of a new personal agenda for clients to replace the old dossier that was kept by professionals only. The new agenda was the basis for discussion between the client and his or her informal care takers, if present, and a team of two professionals, the personnel coach of the client and a care co-ordinator, both chosen by the client. The second exemplar was a new system of learning on the job. This worked as a catalyst to increase both the learning capacities of individual workers in the organisation and enhance cooperation between workers from different disciplines. Accompanied by intensive coaching of middle management and individual workers this exemplar functioned very well in getting demand-driven care started in the organisation. Evaluation took place with middle management, who themselves operated as researchers by preparing this evaluation in their own department.

4.10 The sum of the parts...

On the level of the firm as a whole, of which the different locations are a part, the research has also had its effects. During the whole process close co-operation is established with a staff member of the organisation, who actively takes part in the project and functions as a first class co-researcher. This co-operation creates good conditions for investigating what is general for the organisation, what is specific for the location, and how the two are related. It also makes it possible to see which results from one context can be used in others. The increasing knowledge of the researchers of the different products and processes within the organisation, of the different perspectives of actors involved and of the existing power relations, contributes to the possibilities for generalisation.

4.11 Final remarks

Writing such a short reconstruction of a research project as extensive as this one has serious limitations. Writing means setting out the process in a different sequence and putting things in order that at the time weren't as naturally sequential as it now seems. At times we did not know how to continue and needed lengthy discussions to find a way to make a subsequent step with sufficient basis in the organisation (in this we were not always successful). However, a lack of immediate success in itself also provides new knowledge about the processes under study, although sometimes we wanted things to turn out differently. At times we faced problems in human communication or in power relations that were difficult to overcome. We were sometimes confronted with our own limitations: this kind of research demands a lot of the researcher in terms of knowledge, competences, time and commitment that we couldn't always offer. With the reality under study we had to accommodate the research with the harsh, political and social conditions in which work in this sector has to be done and in which lives have to be lived.

References

Argyris, C., 1990. Overcoming organizational defences. Allyn & Bacon, Boston, Mass.

Argirys, C., 1999. On organizational learning. Blackwell, Oxford.

Beck, U. and E. Beck-Gernsheim, 2002. Individualization. Sage Publications, London.

Beukema, L., 1996. Labour studies as an interventional science: a Dutch case study, Paper presented at the World Meeting of Labor Studies and Sociology of Work: Methodological Concerns, Puerto Rico, February 1996.

Brydon-Miller, M., D. Greenwood and P. Maguire, 2003. Why action research? Action Research 1: 9-28.

Coenen, H., 1989. Handelingsonderzoek als exemplarisch leren. Van Arkel, Utrecht.

Coenen, H., 1996. The model of exemplarian action research. In: Boog, B., H. Coenen, L. Keune and R. Lammerts (eds.) Theory and practice of action research. Tilburg: University Press, Tilburg, the Netherlands, pp. 13-21.

Coenen, H., 1998. Handelingsonderzoek als exemplarisch leren. Een bijdrage aan de fundering van de methodologie van handelingsonderzoek (2e druk). Van Arkel, Groningen/Utrecht, the Netherlands.

Giddens, A., 1990. The consequences of modernity. Polity Press, Cambridge.

Giddens, A., 1991. Modernity and self-identity. Polity Press, Cambridge.

Lennie, J., C. Hatcher and W. Morgan, 2003. Feminist discourses of (dis)empowerment in an action research project involving rural women and communication technologies. Action Research 1: 57-80.

McNiff, J. and J. Whitehead, 2000. Action research in organizations. Routledge, London/New York.

Niessen, J., 1995. Omzien naar een volwaardig leven – de intramurale ouderenzorg vanuit mijn bed bekeken. SWP, Utrecht.

Valkenburg, B. and J. Lind, 2002. Orthodoxy and reflexivity in international comparative analysis. In: Van Berkel, R. and Hornemann Moller, I. (eds.) Active social policies in the EU. Policy Press, Bristol.

Warners, I., 1998. Terug naar de oorsprong. Bohn Stafleu/Van Loghum, Houten, the Netherlands.

5. Preparing scientists for society

A PhD training programme at Wageningen University, the Netherlands

Conny Almekinders, Jet Proost and Jürgen Hagmann

5.1 Scope

In the area of agricultural development, as in many other areas, there is considerable pressure to combine good research with a direct impact in the field. This pressure has generated a call for another type of research, with a different role for researchers (e.g. German and Stroud, 2007). To address farmers' problems, and rural societal needs at large, researchers should break away from the paradigm in which generation of knowledge is seen as an exclusive activity of scientists. This call is voiced particularly strongly on the African continent where the labels like Innovation System approaches (Sanginga *et al.*, 2008) and IAR4D (Integrated Agricultural Research for Development) are attached to it (ICRA/Natura, 2003; Starks *et al.*, 2005). This research is an applied type of research, asking for interdisciplinary approaches to be integrated into frameworks for participatory learning and action research. In this article we refer to such research as 'integrated agricultural research'. Academics that should carry out this research are expected to be effective team players, able to handle complex situations with the involvement of multiple-stakeholders (Chambers, 1997, Patel *et al.*, 2001). While the call for this new type of scientist is fairly loud, less attention has been paid to the competences that researchers should be equipped with (Levin and Martin, 2007) and how these competences can be developed (Patel *et al.*, 2001; Kibwika, 2006). This contribution describes the case of a PhD programme at Wageningen University, the Netherlands, mostly with African students, aiming to prepare PhD students for a role in the new type of agricultural research advocated. The purpose of the multi-year PAU programme (Participatory Approaches in agricultural (technology) development and their Up-scaling) is to investigate how participatory approaches work and how successful experiences can be scaled up. In addition to training the PhD candidates in the conventional academic competences, the programme piloted competence development that emphasised human interaction, communication and collaboration as part of the curriculum. This involved a range of cross-cutting social skills, the related deeper understanding of human (and one's own) behaviour and values, and how these play a role in research work, in particular interdisciplinary research (see also Box 1 and 2).

The implementation of a series of Learning Workshops offered the opportunity to explore the competences for integrated research and how they can be developed. The competences involved far more than social skill training; it is also about mindset and attitude, being conscious of one's personal learning process, built-in reflection in interaction with others, providing and receiving feed back (see Box 1). The experiences in and after the Learning

Box 1. Competences for integrated agricultural research (adapted from Hagmann *et al.*, 2003).

Competences for integrated agricultural research can be considered the product of knowledge, skills and attitude. The skills involve interpersonal skills, academic skills and the skills to apply those in combination in interdisciplinary and transdisciplinary contexts.

- Related to interpersonal interaction (in research):
  - emotional intelligence' in personal development: self-awareness, empathy, critical self-reflection, social interaction;
  - communication;
  - concepts of team dynamics (team functioning, team building, feedback and learning culture);
  - facilitation concepts and practice;
  - negotiation and conflict management (concepts and practice/skills).
- Related to the application of new concepts:
  - process and systems thinking to understand change and how to make change within the context of participatory development;
  - action research: linking theory, practice and learning;
  - interdisciplinarity and transdisciplinarity – in conceptual and team perspective;
  - organisational development;
  - management, planning and design, quality assurance.

Workshops emphasised the importance of inter- and transdisciplinary context in which these interpersonal competencies have to be used. The idea of the 'reflective researcher' was taken on board.

The series of workshops in which the authors were actively involved, also yielded (unexpected) experiences on how such competence development fits into and interacts with an academic environment that holds varying views and expectations regarding PhD research and an appropriate curriculum. This chapter reflects the learning case of the PAU programme and is relevant for academics who are confronted with demands to tailor curricula for the new type of professionals. It is not only of specific relevance for PhD programmes of Northern Universities who train scientists from developing countries, but can be applied in the Northern academic research context as well. And the experiences can easily be expanded to MSc programmes where, for instance, community service learning (see also Bringle and Hatcher, 1995) becomes a valued part of the curriculum.

Box 2. Learning Workshops.

The First Learning Workshop 'Facilitating change in upscaling of participatory approaches' (2002)

The PhD candidates from the 1st and 2nd cohort of the PAU programme participated, together with three other participants, also with scholarships from Rockefeller Foundation, in the first off-campus workshop. Two trainers/facilitators (3rd author) and the coordinator of the PhD programme (1st author) were responsible for the design and implementation, and documentation. The Rockefeller Foundation financed the workshop.

In the 9-day workshop a combination of methods was used, addressing both the content of the PhD research programme and competencies building needed for this type of research. Through various monitoring committees the participants took ownership for the training process. The two trainers/facilitators implementing the workshop followed the priorities of the PhD students in terms of skills they liked to be introduced to and practise, and topics they wanted to further elaborate. They gave input by presenting concepts and theory. The underlying idea was that research on participatory approaches in agricultural research and development would require a range of skills and insights related to personal qualities and interactive communication: listening, building team relations, facilitation group discussions, etc. These are cross-cutting social skills, related to empathy, emotions, beliefs and elements of personal character. To get insight into one's own performance and skills different role plays, exercises and reflection sessions were used, followed by practising ways in which to improve one's own performance. These skill-based competencies were woven through a programme that also addressed the research issues in the programme and the individual PhD projects. The trainers/facilitators, being involved in several institutional change processes in Africa themselves, could present many illustrative examples from their own work. This helped participants identify the relationship with their own research context.

All techniques used in the workshop were combined in a 'toolbox report' and the entire workshop process/design was documented in a report.

- Content of the PhD programme.
  - SWOT analysis of the PhD programme.
  - Envisaging the outcome of the programme and PhD research projects.
  - Presentation of research cases: how to study change, how to use research information for bringing about change, action research.
- Competencies.
  - Getting to know yourself: reflection on values and beliefs.
  - Team work: personalities and team roles, origins of tension, how to improve team work, provide feedback.

- System Change, bringing about change in human systems and responses to change.
- Analysis of personalities and styles.
- Methods/tools used.
  - Group work and plenary brainstorming and discussions.
  - Role plays combined with theory: observing and interpretation, experiential learning.
  - Facilitation: principles and practice (through group work and discussion).
  - The art of questioning: interview techniques.
  - Visualisation.

The following Learning Workshops had similar set ups, incorporating the lessons learnt from earlier workshops. They differed in length, composition of facilitation-training team and students participating.

Second Learning Workshop in 2003 targeted the 3rd cohort of the PhD programme. It was a 5-day workshop, also at a venue out in the Dutch countryside. The same consultant-trainer (3rd author) was involved. This time, however, the design and implementation was a joint effort with the programme coordinator (1st author) and four of the PhD students who participated in the first Learning Workshop, a year before. In this way they could practise and further develop their competences. The participants were again foreign PhD students from Wageningen University, coming from one chair group; the 3rd cohort of the PAU PhD programme was the core of this group.

Third Learning Workshop in 2004 targeted the 4th cohort of the PAU PhD programme. It was also a 5-day workshop, at the same venue as the third. This time the trainer/facilitator (3rd author) teamed up to design and implement the programme with the programme coordinator (1st author) and another facilitator/trainer (2nd author) who also worked as university staff member. This workshop was attended by a combination of PhD students from various PhD programmes and chair groups, foreign and Dutch. In addition, two Dutch MSc students in development studies participated.

5.2 The academic curriculum for agricultural development-oriented research in Africa

After the realisation that the Green Revolution of the '60-70s did not bring the expected improvement in the livelihood of small-scale farmers in developing countries, participatory approaches became the advocated way of working in development-oriented agricultural research. The 'participation' paradigm called for researchers to work directly with farmers in farmers' fields and deal with their problems. Farmers' collaboration in experiments and their expertise about the local situation transformed the farmers from recipients of technologies into co-researchers. A large number of different forms of participatory research have since emerged, e.g. Participatory Extension Approach (PEA), Participatory Learning and Action Research (PLAR), Farmer Field Schools (FFS), and Local Agricultural Research Committees (LARCs). All these forms contain elements of action research; they combine research, learning and action in different forms. Increasingly, with the move from farmer participatory research into system innovation, other stakeholders such as service providers and intermediaries became part of the research processes as well.

The participation of farmers and other stakeholders also brought along the use of a broad range of interactive techniques that are relatively new in the arena of agricultural research (focus group discussions, transect walks, participatory rural appraisals, ranking and rating exercises, etc. (see Pretty *et al.*, 1995)). The mode of working in participatory research has become pre-dominantly group-based. Knowing how to be a team-player, organise workshops and facilitating group-based discussions have become essential skills for development-oriented agricultural researchers in Africa.

Many feel that universities are currently not focusing sufficiently on the competences that researchers need in order to play their new role in research (Patel *et al.*, 2001; Kibwika, 2006). Universities traditionally excel in research training for scientific knowledge production – in many instances within a paradigm that considers knowledge production an exclusive domain of scientists (Gibbons *et al.*, 1994). In this paradigm, generating disciplinary knowledge through data collection and analysis, reasoning and theorising are the core competences of a scientist. In such a paradigm, there is no space in an academic curriculum for competences which relate to skills for human interaction. However, there are also many examples of groups in universities which deliberately engage in interdisciplinary research, aiming to bridge the gap between generating scientific knowledge and societal impact. Several PhD programmes at Wageningen University seek the integration of natural and social sciences to address complex problems. This type of research often implies that the scientist is 'embedded' in on-going change projects and processes. Many researchers – like those participating in the PAU PhD programme described in this chapter - , are thus participating in some form of action research, like open-ended collaborative experimentation with agricultural technologies (see for example Ramaru *et al.*, Chapter 3), but also in processes particularly focusing on social change (for example Mutimukuri, 2005). Therefore it is very important to make students understand the competences needed to implement such research. The following sections elaborate on the experiences and learning from the PAU PhD programme, which provided a protected space for developing an unconventional module in the academic curriculum.

5.3 Background of the PAU programme and the university context

In the year 2000, a Chair group in the Social Science department of Wageningen University, Technology and Agrarian Development (TAD) obtained financial support from the Rockefeller Foundation for a special PhD programme. Eventually this resulted in grants for 24 sandwich scholarships and special activities. TAD recruited for the programme over a period of four years (2001-2004), four cohorts of 6 PhD researchers each. The majority of the recruited PhD candidates were researchers with a natural science background from international and national agricultural research institutions and universities in East and Southern Africa (see Table 1). Most of the organisations involved receive(d) support from the Rockefeller Foundation, among others, for participatory technology development with

Table 1. Back ground of PhD candidates of cohort 1, 2, 3 and 4 of the PAU programme (compiled by I. Ruisch).

Institutional affiliation
Cohort 1
• Lecturer, Dept Extension, Makerere University, Kampala, Uganda
• Researcher, Network for Rural Development Studies (NGO), Mexico
• Lecturer, Dept Extension, Makerere University, Kampala, Uganda
• Senior researcher, Kenya Agr.Research Institute, Nairobi, Kenya
• Researcher Multiple Cropping Centre Chiang Mai Univ, Thailand
• Researcher, CIFOR, Harare Zimbabwe
Cohort 2
• Lecturer, Dept. Animal Sciences, Mekelle University, Mekelle, Ethiopia
• Researcher, CIMMYT/Bunda College of Agriculture, Lilongwe, Malawi
• Lecturer, Dept Extension, Makerere University, Kampala, Uganda
• Lecturer, Limpopo Dept. of Agriculture, Limpopo, SA
• Researcher, African Highland Initiative, Kampala, Uganda
• Researcher, KARI, Embu, Kenya
Cohort 3
• Researcher NARO, Kampala, Uganda
• Researcher, Kenya Forestry Research, Nairobi, Kenya
• Researcher, TSBF-CIAT, Nairobi,Kenya
• Staff member, Coast Development Authority, Mombassa, Kenya
• Research Fellow, Institute for Development Studies of Nairobi, Kenya
• Researcher, National Institute for Soils and Fertilisers, Hanoi, Vietnam
Cohort 4
• Researcher, Industrial Ecology Institute (NGO), Nairobi, Kenya
• Researcher, WARDA, Cotonou, Benin
• Lecturer, Co-operatieve College, Moshi, Tanzania
• Researcher, KARI, Nairobi, Kenya
• Researcher, Limpopo, Dept. of Agriculture, Limpopo, SA
• Researcher, Honeybee Network/National Innovation Foundation, India

and for small-scale farmers in the region. The linkage with these research institutions would allow the programme, through PhD training and research, to contribute directly to capacity development in the home organisations of the PhD researchers and the rural communities they were working with.

Previous academic training	Year of birth
MSc Ag. Extension and Education, Makerere University, Uganda	1968
MSc Agronomy, Colegion Postgraduados, Montecillo, Mexico	1972
MSc Ag. Extension and Education, Makerere University, Uganda	1973
MSc Agronomy, Texas University, USA	1957
MSc Agriculture, Chian May University, Thailand	1970
MSc Agricultural Knowledge Systems, Wageningen University	1975
MSc Animal Sciences, Wageningen University	1971
MSc Agroforestry, Bunda College, Malawi	1965
MSc Agricultural Extension and Education, Makerere University	1965
MSc Home Economics, Univerity of Christian Higher Education, Potchefstroom	1972
MSc Arts in Anthropology, University of Nairobi, Kenya	1966
MSc Agricultural Economics, University of Nairobi, Kenya	1963
MSc Development Studies, University of East Anglia, UK	1970
MSc Environmental Forestry, University of Wales, UK	1968
MSc Anthropology, University of Nairobi	1973
MSc Agricultural Knowledge Systems, Wageningen University	1960
MPhil. Environmental Studies, Moi University, Kenya	1966
MSc Agriculture, Hanoi Agricultural University, Vietnam	1974
MSc Arts, University of Amsterdam	1971
MSc Crop Science, University of Benin	1960
MSc Arts – Women and Development, Institute of Social Studies, The Hague	1958
MSc Arts – Urban and Regional Planning, University of Nairobi	1965
MSc Soil Science, University of Reading, UK	1962
MSc Plant Pathology, Ravishankar University, Raipur, India	1963

The research theme and PhD candidates

The thematic scope of the PhD programme was 'Participatory Approaches in agricultural (technology) development and their Up-scaling' (PAU). Since the Rockefeller Foundation had been supporting many projects and institutions in the development and application of participatory approaches in agricultural technology development, they were interested in understanding which participatory approaches worked best and how they could most effectively be scaled-up, to reach more farmers in a shorter period of time with the benefits of agricultural research. The research in the PhD programme would be instrumental in giving answers to those questions. Most recruited researchers had some years of professional experience in projects that applied participatory approaches. TAD staff felt that these PhD researchers would bring in the real experiences from the on-the-ground participatory work and the agronomic background of many of the PhD candidates could be an asset in the studying of this theme. The theme of study fitted very well in the core interest of the TAD chair group which finds itself at the intersection of science and technology studies and development.

The tailor-made curriculum

The recruited PhD candidates were provided with a sandwich scholarship which consisted of a 10-month preparation programme in Wageningen University and another 10-month period to finalise the writing of the thesis. The field research was to be carried out in their home country during the 2-year in-between period (Box 3).

The 10-month preparation programme at Wageningen University was intended to support the PhD candidate in the development of a full research proposal and strengthen the weaker competences of the candidates. Normally, PhD candidates pick and choose from existing international MSc programmes, special PhD courses, seminars and workshops that are offered during the period a PhD student happens to be in the University. Having cohorts of 6 PhD students with a similar time schedule offered TAD opportunities to 'tailor' a PhD programme through the organisation of special workshops and courses.

The idea of organising a special workshop to address the competence development for participatory research, stemmed from the perception of the PhD candidates and programme coordinator (first author of this chapter) that the training programme should make PhD students more effective professionals, competent to carry out a better job in the field. For them doing a better job meant 'working (i.e. 'participating') more effectively with farmers'. Participatory research with farmers was for most of them part of their job at home and presumably their PhD research would involve similar work. Thus, as PhD researchers they would not only study participatory approaches as a subject, but many of them would also be part of it as member of a project or research team. Since they knew Wageningen University as one of the most respected universities in the world, they expected this to be the place

Box 3. Structure and characteristics of the PAU Phd programme (2001-2008).

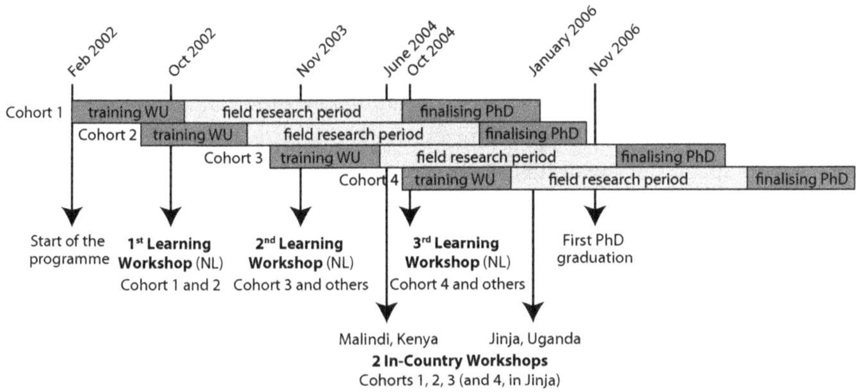

The programme recruited 4 cohorts of 6PhD candidates each who started in February 2002 (cohort 1), September 2002 (cohort 2), September 2003 (cohort 3) and September 2004 (cohort 4). Each cohort followed a 4 year plan: 10-month training period in WU, followed by app. 2 years of field research and a final period of PhD writing in WUR. Three Learning Workshops were organised in Netherlands, and 2 In-country workshops to support the PhDers in their competence development and progress research work. The first PhD candidate graduated in November 2006.

where they could get the best training in doing participatory research. Within the regular PhD training programme of the University there was no module or provision that seemed to meet this expectation, and therefore the obvious option was to organise such training within the context of the PAU programme. The Rockefeller Foundation provided an extra grant for this purpose with which the programme coordinator (1st author) contracted a trainer/facilitator (3rd author) to design and implement an action learning module. This resulted in the first Learning Workshop, in the fall of 2002, when the 12 PhD students from both the first and second cohort were both present in the Netherlands. Later more workshops followed, in which the 2nd author became involved as a trainer/facilitator.

5.4 The Learning Workshops and their context

The First Learning Workshop

The motto of the first Learning Workshop was *to know yourself as a basis for understanding others*. During the 9-day workshop the PhD students reflected on the aims of the PhD research and their participation in the PAU programme, and to what extent their research would or could contribute to societal change. They were encouraged to reflect on the importance of human behaviour in team work, institutional change, collaborating with communities, at the same time understanding their own behaviour in these situations through exercises, group work, and plenary discussions. A central element was the question as to what participation and change means in reality and in research. Participants were stimulated to challenge their perceptions and realities, and their behavioural patterns around participation and change. Throughout the programme methods and techniques were used that provided participants with opportunities to practice skills and understand underlying concepts that are specifically associated with effective team work, collaboration with farmers and personal reflection and feedback (See also Box 3). An output of the workshop was a group vision of what the PhD programme should entail. They saw how a research process can be appreciated as a learning process.

This Learning Workshop was highly appreciated by the PhD students and the programme coordinator (1st author), as was clear from the discussions and evaluation. Examples and role plays were taken from situations participants were familiar with, which increased the impact of the sessions. Participants felt they had been filled-in on issues and meanings of 'participation' and their research; on how to do things differently. They had also been practising doing things differently, for example, giving each other feed back and facilitating group sessions. The workshop experience motivated them even more to ensure that colleagues and farmers back home would directly benefit from their PhD project. The exchange of personal experiences, values and beliefs had felt like a 'warm bath' after having been in the university for some months, in which lectures and reading had been their central activity. Sitting for long hours in their student room or office, away from their family, during a Dutch winter had not been encouraging either. When the PhD students returned after the workshop to daily life in the university to resume the regular course work, they were even more critical about the need to read up on the big social science theories. These texts were difficult to read for them, using vocabulary and frames of thinking with which they were not familiar. Their main interest was in studying human behaviour and the implications of change.

The differences with their own academic culture became apparent as well. Where in most African universities professors hold the truth and power over decisions regarding the PhD work, in the Dutch university system PhD students were considered scholars and were challenged in discussions with staff. Instead of feeling challenged however, they felt at times intimidated or not taken seriously in their new learning. In addition, finding their way in the administrative and educational system – which may have been more straightforward for Dutch students - was challenging as well. For example, not only did they have to define and write their PhD research proposal, they were also expected to find themselves a supervisor whom they considered most appropriate. At the time, this added to the struggle of understanding social sciences. In retrospect we can say these struggles have helped PhD students to realise that it was not their own weakness, but apparently part of the difficulty of understanding the system and becoming an interdisciplinary type of scientist.

An important factor was that other staff members had little information on the content of the Learning Workshop and the expectations of the PhD students. Some staff members emphasised that PhD training means to prepare PhD candidates in the first place on 'how to research participatory approaches', the purpose of the PAU programme. The recruited PhD students obviously needed to strengthen their competences to carry out good scientific research; they therefore needed a lot of scientific reading, training of their skills in theorising, reasoning, conceptualisation and synthesis. Everyone agreed that although the PhD candidates were becoming interdisciplinary hybrids rather than genuine social scientists, they were after all doing a PhD. And, this implied that they would eventually have to successfully defend a PhD thesis, which is essentially proof of academic research competences. Some TAD staff considered that the focus on competence training as in our first Learning Workshop came at the expense of time spent on literature study and proposal writing. The content was seen as relevant, but something that should stay outside the academic curriculum and be picked up elsewhere in life or from other training institutions. In addition, there was doubt about the effectiveness of such training for doing better PhD research: there was (still) no evidence to prove for it. The views of the staff regarding roles of scientists and development-practitioners were not formulated explicitly, but students and the programme coordinator overheard and interpreted loosely made comments over lunch or other informal events. For instance, the professor jokingly used the term 'bean-bag-throwers' when he referred to the professionals working in the development scene.[8] From the students' perspective, this showed the professor's aversion to this world of professionals of which they felt part. Several PhD students, for their part, felt that quite a number of academics did not know 'the reality on the ground' and provokingly used the term 'ivory-tower-dwellers' to label university staff. Comments and incidents confirmed the lingering impressions of students and programme coordinator and blocked them from freely discussing their views and ambitions.

Conversations between the PhD students and the programme coordinator showed a growing uneasiness with the situation from the point of view of the PhD students and the programme coordinator. The question 'but who will milk the cow?' became metaphorical for their concern about the usefulness of so much theoretical reading for solving the problems of the African farmer. How could reading the work of Durkheim and other social scientists change the functioning of their institutions, help them in their ambition to participate better with farmers and have impact on farmers' livelihood in their home country? And what would happen if they wanted 'to do action research' – which was so much advocated in circles they were part of back home; what were the consequences if they did not design and frame their research the way the professor wanted? These perceptions and frustrations persisted when the PhD students started their field research in their home countries. In a meeting with the programme coordinator and representatives of the Rockefeller Foundation in Uganda, June 2003, the PhD researchers vented their concerns about the alienating

[8] Based on the professors' notion; those professionals often hold workshops in which participants frequently throw around a bean-bag – as a way of energizing and breaking the monotony of a meeting.

theoretical orientation of the programme. The question as to whether the programme could 'blossom' and achieve its objectives in the current academic environment became a point of discussion. The representatives from the Rockefeller Foundation felt that the PhD students had a valid argument. Their involvement in the agricultural research and development scene in Africa showed them the need for a new type of professional. The programme coordinator (1st author) advocated that this new type of professional would not be able to purposefully engage in the action if they could not at the same time be critical analytical researchers, able to interpret the situations they found themselves in. To this end, theoretical understandings of the issues at stake are a necessary component of research training.

The third and fourth cohorts and their Learning Workshops

While the PhD students of the 1st and 2nd cohort were doing field research in their home country, the newly recruited PhD candidates of the 3rd and 4th cohort started their PhD training in Wageningen in September 2003 and September 2004, respectively. The curriculum of both cohorts was similar to that of the first two, except for the fact that these latter cohorts were not present at the same time in Wageningen (Box 3). The PhD students also took courses from the regular PhD and MSc programmes, read up on literature, worked and interacted on the development of their research proposals. For each of the two cohorts the programme coordinator (1st author) organised learning workshops similar to the one described above (Learning Workshops 2 and 3, respectively, see Box 1 and 2), now also inviting PhD and MSc students from other programmes with an interest to develop their competences for integrated agricultural research. The two Learning Workshops for the 3rd and the 4th cohort were considerably shorter than the first one (41/2 and 5 days respectively) because of limited financial resources and supervisors of the participating students considered that 9 days was taking too much time from the students' programmes.

Because the 3rd and 4th cohort of PhD candidates went through a training programme comparable to the first two cohorts, experiences with the Learning Workshops and reactions of the environment were more or less similar. The most salient observation was the absence of signs of tension in the relationship between PhD candidates and staff. The PhD students from the 3rd and 4th cohort seemed more appreciative about the studying of social science theories and the application of them in their research proposal than the students from the first two cohorts. In addition, the TAD staff seemed more open to the experiences of PhD students, including their experiences in the Learning Workshops. There was a more relaxed situation with more interaction between PhD students and staff members. The input of staff members in response to queries and draft research proposals were important stimulations.

Comparing experiences between cohorts

How could the absence of conflicting expectations be explained? Had the experiences with the first two cohorts helped to improve the training period of these two subsequent cohorts? The situation of the last two cohorts of PhD students differed from those of the first two on several points:

- First of all, the recruited students from the 3rd and 4th cohort had a somewhat different background. Whereas several of the PhD candidates from the 1st and 2nd cohort were involved in ambitious institutional-change projects, in which the process consultant (3rd author) was involved as well, the PhD candidates of the 3rd and 4th cohort included several persons with training in social sciences (see Table 1). This may have cushioned the shock in the cohort of being exposed to social sciences literature and theoretical reflection.
- Programme coordinator and professor were better prepared for the expectations of the next PhD candidates: the curriculum with proposal development, reading and course work was better structured, and theory was presented to the PhD students in a more comprehensive way. Attention was given to bridging the usefulness of having a theoretical framework for data collection and interpretation and the nature of their PhD research work.
- The coordinator took time to talk with the students about the purpose of a PhD, Dutch academic culture, and university organisation to avoid misunderstandings and false expectations with talks about Dutch academic culture. She explained to the students that other staff members of the group were interested in the research plans of PhD candidates, but that staff could not always help with their proposal presentations because of their high workload.
- The programme coordinator made efforts to explain to PhD students and university staff her view on the relevance of the Learning Workshops for PhD research. She pointed out that in the modules a broad range of learning activities was offered, to accommodate different learning styles and multiple intelligences. Students explored their own styles and reflected on their own behaviour. At the same time students trained their skills of observation and interpretation of body language, group dynamics, etc. She made it clear to staff members that in the workshops students would also practise interviewing skills by means of a range of assignments. Particular concepts described in behavioural models were illustrated by experiencing reactions and emotions in the role plays and through assignments. It was also emphasised that action research can take the form of 'embedded research' and is not necessarily 'activist research'.
- The two following Learning Workshops were shorter and included students from other programmes. This reduced the cohort feeling, i.e. the 'bonding' effect within the group, but at the same time reduced a potential feeling of frustration and alienation from the university environment.
- A plan was made for the workshop participants to share their experiences with the professor (3rd cohort) and a wider circle of university staff (4th cohort, see Box 4). The appreciation of the PhD researchers for what they had experienced in the workshop and the importance it had for them as people and as researchers was undeniable and hard to ignore.

Follow-up workshops

A follow-up workshop was organised in June 2004 in a hotel in Malindi, Kenya for the

Box 4. Bringing the learning to a wider academic audience: Gallery Walk.

The programme coordinator was concerned about defiant comments from university colleagues reluctant to accept the Learning Workshop as an essential part of the PhD curriculum. Toward the end of the 2nd Learning Workshop, the 1st and 3rd authors in their roles as programme coordinator and trainer discussed with the participants how they could improve the understanding among university staff about the Learning Workshop. The participants decided to organise a meeting with the professor, to brief him and share their views on the importance of this workshop for their PhD process.

In the 3rd Learning Workshop the participants themselves came up with the idea of informing others. They wanted to make the university see the importance of a competence training of the kind they were experiencing. For this purpose, the trainers organised what they called a 'Gallery Walk' towards the end of the workshop, All posters and other visualised products from the workshop were plastered on the wall, following the sequence of the programme they had gone through. By revisiting posters and visuals participants got a good overview of the entire workshop programme. This deepened their understanding of the interconnection and meaning of the course elements. Returning from the venue of the Learning Workshop, the PhD students organised a similar Gallery Walk in the University. They invited supervisors and other important people they wanted to inform about the competence training. The students guided the audience through the poster series, indicating how the various products had come about, how they related to each other and what the importance was for them. The posters included the ones in which they had documented their own evaluation of the workshop (see below). Each Learning Workshop was evaluated by the participants. Below are some of the participants' comments on the third Learning Workshop.

If I had to explain to my supervisor the importance and role of such a workshop, I would say...

- That it made me realise the importance of thinking/looking at research from an *impact*-oriented approach. And that I learned what is needed (skills) to conduct such research
- I would like to do a PhD that has/brings life (making a difference to the system I am coming from)

- This workshop changed something in me and made me more oriented towards field research
- Ask if she/he is able to put her/his head out of his/her body (heart and imagination)
- I will think in another definition of an holistic approach
- Development of my competence
- Integration of one's personal & professional goals in a PhD
- Better to develop both sides of the brains than developing only one side
- The outcome of research could make a difference to society, who think and solve problems in their own way
- That the most important thing to consider during the PhD work is how it impacts (anchor)
- Relating personal development & impact
- Personal development + relevance of research
- A different view on reaching impact
- The workshop was important and that many people after their PhDs do not make it in life because they lack the skills I learnt.
- A new perspective on system theory
- I want to include all the human messy part in my thesis

Photos: (left) participants filling in evaluation posters; (right) PhD students present their workshop experiences in the form of a Gallery Walk to interested University staff.

PhD candidates in the first three cohorts and a number of supervisors. They came together to exchange experiences and discuss progress in the fieldwork. The mode of working in this follow-up workshop was similar to earlier workshops: the same trainer/facilitator (3rd author) helped to design the workshop programme in such a way that the participants took active part and practised some of the skills introduced in the Learning Workshops. This follow-up workshop was also the first occasion for PhD students of the 3rd cohort to meet with PhD students of the first two cohorts. The start of the workshop turned out to be a landmark in the programme. There was tension among PhD students of the 1st and 2nd cohort. Some of them felt the same frustrations from the Wageningen period coming back. They were not confident about the appreciation and respect from the supervisors. The lack of trust was felt more strongly now because they had their first field data. Could they freely discuss their findings and ideas? Stories about professors stealing their students' findings were common in their universities. The 3rd cohort seemed unhindered by such feelings: they had no field data to protect and also had no reason to distrust intentions of supervisors. The facilitator decided to throw it onto the table. A serious discussion flared up, which led the professor, sitting in front of the group to publicly announce that whatever the orientation of the thesis was, he would only judge it on its 'academic defensibility'. Even if there was an approach or topic he could not be sympathetic with, he would lead the PhD candidate to promotion. This public statement re-assured the students and removed the tension that a couple of them struggled with. Since then, differences of opinion have remained, but have now become a topic of fruitful academic discussion, not hindering the finalisation of PhD theses in the programme.

In January 2005, a second follow-up workshop was organised in Jinja, Uganda. In this workshop all four cohorts participated and a set-up similar to that in Malindi was used. For PhD researchers it was an important event to reflect on the progress of their research, identifying challenges and exploring difficulties in consultation with colleagues and friends. The fact that participants were at different phases in their PhD research and their challenges varied accordingly resulted in valuable exchanges of experiences among the PhD students.

Last phase of the PhD research

After their fieldwork PhD researchers returned to Wageningen University for the final writing up of the thesis. PhD students returned to Wageningen University at different times, depending on the cohort and the individual progress of the thesis work. Consequently, the composition of the group of PhD students present in Wageningen varied as well: PhD students from the first cohort were now sharing offices at the university with those of 2nd, 3rd and even 4th, etc. In this last phase of the PhD, students struggled to sort their information, interpret findings and write thesis chapters. Taking a critical stance on the situation in which they had been working and collecting data was a challenge. Also, to develop a way of writing in which empirical data are the foundation for any conclusion appeared to be rather difficult for some of the students who had become very skilled in progress report writing for donors. For the integration of theory and empirical data, and developing conclusions, they now relied on conventional academic skills. Appreciation for more theoretically-oriented discussion and literature grew – this was the only way they could 'make sense out of their findings'. Although we expected peer learning after the sequence of workshops to continue, interaction among PhD students diminished: they did not want to waste time on discussing each others' findings, or going out together, or on other activities which distracted them from their own individual piece of work. They concentrated only on finishing within time.

5.5 Reflection on the experiences

Looking back and analysing our experiences, we can say that we have learned from the competence development initiative in two ways: how to develop a curriculum that addresses competences for integrated research and how piloting something new interacts with the institutional academic environment.

Learning from the learning

At the level of skills training, offered in the workshop, it was obvious from the evaluations of the workshop participants that there is a demand for such 'learning' (see Box 1 and 2). The Learning Workshops served the participants in different ways. For the majority of the participants the workshops were eye-openers which they considered important for their research and personal life. The eye-openers related to the entire range of topics in the workshop programmes: communication- and behavioural models, one's own strengths and weaknesses in interaction with others, ways to continue developing one's skills, reflections on one's own research and theoretical concepts like process and systems approach, learning, action research.

In addition, several PhD graduates referred in their final briefings after graduation, to the workshops as the most relevant activity in their PhD training. At least three PhD graduates applied elements from the workshops in their research with farmers. Several others have mentioned that what they learned about themselves and about human behaviour in general has helped them in their personal as well as in their professional life. In addition, several PhD students have adapted their research or even completely changed the research topic because of the learning and inspiration they got from the workshops. Most of the PhD research in the programme is interdisciplinary and quite a number of students deal with action research or research that critically reflects on the projects and institutions of which they themselves are part. Two PhD studies specifically include the study of interpersonal and communication skills development for research and extension. Several graduates are now regularly invited as professionals in the organisation and facilitations of workshops.

The experiences with the Learning Workshops do not yet tell us enough about the most effective way to develop the skills we are talking about. A four- to five-day Learning Workshop is a good start for developing the competences involved, but more time and effort is needed. Such a workshop can only help to open Pandora's box: show the participants what kind of skills are involved, the concepts behind them and how they can possibly improve their competence in the various areas. A workshop can be helpful in reflecting on the candidate's ambitions in terms of the impact (s)he strives for with the PhD research and explore the opportunities to accommodate these ambitions in setting up the research work. The consequences of choices in terms of competences that are needed to successfully carry out the research can be discussed. The longer, 9-day version of the Learning Workshop could offer more time to practise the skills, deepen insight and further develop the competences the PhD students were interested in. In the 4-5 day version, there was only a brief time to practise and indicate how one could further develop the competences.

The development of interactive skills was confined to special workshops, separated from the course work in which the more conventional academic skills were central. This has not been helpful in showing the complementarities of the various types of skills and related conceptual views involved. Because of the separation, and the focus on what they needed in field research, the interpersonal skill element dropped out of sight later when analysis and writing required the PhD researcher to work in a highly individualistic mode. Finding an effective way of integrating interactive capabilities development into the more conventional academic skill training could also help in an understanding of particular social theories and the usefulness for PhD research work. Experiential learning in role plays based on the students' cases proved useful in this sense. Also the drawing of parallels between what took place in the Learning Workshops, at the level of the PhD programme in the university, in their own research institution and at community level provided useful entry points for showing the usefulness of social theory in formulating research questions and research process design, planning of field work and interpreting situations.

Learning from the institutional environment

The experiences also showed us some of the pitfalls of pilot projects and change within institutions. In our case, the project created a protected space for trying out new things: the project was brought in with external funding and could therefore operate fairly independently. The TAD group discussed and planned the orientation of the programme, but the meanings of the 'tailor-made' element in the programme and differences in underlying visions and goals did not surface until the activities were implemented. As a result, the Learning Workshops were crucial for everybody to understand that concepts such as tailor-made curricula and integrated or impact-oriented research can be taken in different ways. The emerging different views created tensions which in turn led to another dilemma. On the one hand the project coordinator and PhD students needed a protected space to freely experiment with new competence development, unhindered by criticism. On the other hand, there was a need to share ideas and experiences with the wider environment. The high workload among staff members was, however, a factor which did not favour the exchange of ideas: busy daily programmes left little space for focusing attention on innovative things.

Another point of consideration is the importance of the background and expectations of the PhD candidates. In the PAU programme all PhD candidates were from developing countries, most of them from Africa, with training in crop or soil management, plant production and development of rural communities. They were used to hierarchical organisational cultures and academic cultures that tend to be very static. In addition, many of them had a background in natural sciences and were used to seeing social sciences seriously criticised in their environment for their supposed lack of relevance and impact. While not suggesting that only and all African universities are hierarchical and follow the conventional 'scientific' paradigm, this background was not helpful to the PhD students for interpreting and positioning themselves in the environment of Wageningen University.

Apart from the differences in expectations of the parties involved, the reluctance to change has probably played a role. The university, perhaps even more than other types of institutions, is rigid in the face of change: without evidence of usefulness it tends to be difficult to convince scientists of the need for new approaches (see for instance Bawden, 1995). When pilot projects are implemented for the purpose of providing experiences, and testing new concepts in the curriculum, it is desirable to involve people in leadership positions, in order to champion the ideas and experiences, create space and ownership for continuation or wider application. This insight – the need to involve crucial institutional players - has been particularly useful in shaping the strategic planning and initiative so that it incorporates the kind of competence development for IAR4D type of researchers in the curriculum of Makerere University, Uganda (Kibwika, 2006, Hagmann *et al.*, 2007).

It is worth noting that another four-day Learning Workshop was implemented in 2007 with a group of PhD candidates of predominantly Asian origin with financial support from Wageningen University. The programme was adapted, on the basis of earlier experiences, to connect more closely with the PhD research and the type of interactions that the PhD candidates expect to deal with. A new Learning Workshop is planned for the 2008/2009 season, again with financial support from Wageningen Graduate School.

Conclusions

When we started out with the objective to fill a gap in the PhD programme and address the more interpersonal competence development for a integrative researchers' profile, we placed a great deal of emphasis on relational competences, visions and values, on 'knowing what' (the concepts and theories) in combination with 'knowing how' (the practising of interaction and reflection). These are without doubt the essential ingredients of a reflexive researcher. However, as we gained experience, not only with the Learning Workshops, but also with the competences of the PhD students in doing their actual research work, including the writing-up phase, it became clear that these interpersonal competences are very much intertwined with the academic competences.

The ability to relate and communicate is crucial in the application of specific research tools like interviewing, facilitation of group discussions, etc. But there is more. To be able to analyse, conceptualise, and discuss content matter these competences are most relevant in the formulation of (interdisciplinary) research questions, planning of a research process with others, and communicate findings to other audiences. The new type of professional mentioned at the start of this chapter needs to have the ability to handle different types of knowledge, frameworks of thinking and patterns of interaction. (S)he is capable of selecting the appropriate research methodologies related to purpose of study and context. This ability asks even more from a researcher when dealing with other scientific disciplines and with other (non-scientific) audiences. The reflexive researcher integrates it all: personal social skills, disciplinary knowledge and skills, and the translation and application of these in the interdisciplinary and transdisciplinary research field. The learning case described in this chapter serves as an interesting example of the way in which these different competences come together. The development and implementation of the pilot project, but also the interpretation of the unfolding process in the university environment asks for a combination of soft and hard academic skills also on our part, coordinator and trainers/ facilitators: personal characters and relational skills, knowledge about research methods and methodological stances in research, all in practice and in theory, were relevant in an integrated way.

Not all questions raised in relation to the development of the competences that researchers need to operate in interdisciplinary situations can be fully answered. We learn as we go, and integrate our learning in the concept of the Learning Workshops for PhD students. More PhD projects have to be completed first, before we can track the usefulness of the concept. Furthermore, the way the various competences can be integrated and trained is a challenge which we need to address more specifically. For example, addressing personal relational skills in isolation seems a suboptimal option. The relationships a researcher has to deal with go beyond daily-life interactions. It involves relationships with local and university supervisors, colleagues in interdisciplinary research teams, stakeholders in projects or communities. It also involves communication about disciplinary knowledge and theories, construction of shared meanings of problems and solutions with colleague-scientists as well as non-scientists. For that reason, the interpersonal skills should be closely intertwined with the scientific paradigms and knowledge frameworks used by different actors. Considering the need for integration has implications on the way training is organised. For example, in relation to the question about whether participants should preferably be from similar colliding disciplinary fields, i.e. working in the same research arenas, or not. While we see the disadvantage of having confined these skill elements to workshops, we do not want to advocate that all PhD researchers should embark on the same type of research. In relation, not all students will feel the same need to develop their expertise in group dynamics, process management, etc. To introduce specific competences in a Learning Workshop mode is therefore a pragmatic option for the moment. Interested students can sign up and workshops can be tailored to the specific research domains and background of the students in order to achieve maximum impact. In any case, we consider that a university that takes interdisciplinary and research-for-impact on board as important concepts in its research programme should have provisions in its curriculum for those who want to engage in such research.

5.6 Note from the authors

The PhD programme was an open-ended initiative and not set up as an action research project. When the programme started, it did not have the objective to change the PhD curriculum or the way to go about things. Neither did it have the prior intention to turn the PhD students into reflexive researchers. The experimental character of the programme, as it unfolded, did however invite reflection. We have not wanted to prove anything but wanted to make sense out of what we experienced and draw useful lessons. The programme had a 'participatory' character in the sense that PhD candidates had an important say about their own learning and the programme activities and orientation. Their comments and reflections on the experiences have been captured by the programme coordinator through individual interviews or as they emerged in discussions and workshop sessions. At the moment of submitting this text all PhD students and staff members have been invited to read the draft text of this contribution. Based on their reactions text has been added or altered on some points. We hope this provisional final version will generate even more learning from the process we have experienced together.

Acknowledgements

The authors would like to thank all who contributed to this article by reading and responding. The comments of Paul Richards were appreciated and useful for improving the text. The support from Leni Beukema and Coyan Tromp was essential for the reflection on our experiences.

References

Bawden, R., 1995. Systemic development: a learning approach to change. Occasional Paper no. 1, Western Sydney Univ. Centre for Systemic Development, Australia.

Bringle, R.G. and J.A. Hatcher, 1995. A service-learning curriculum for faculty. Michigan Journal of Community Service Learning 2: 112-122.

Chambers, R., 1997. Whose reality really counts. Putting the first last. Intermediate Technology Development Group, London.

German, L., and A. Stroud, 2007. A framework for the integration of diverse learning approaches: operationalizing agricultural research and development (R&D) linkages in Eastern Africa. World Development 35: 792-814.

Gibbons, M., C. Limoges, H. Nowotny, S. Schwarzman, P. Scott and M. Trow, 1994. The new production of knowledge: The dynamics of science and research in contemporary societies. Sage, London.

Hagmann, J., P. Kibwika and A. Ekwamu, 2007. Learning to make change. Developing innovation & change competence in African Universities. Paper presented at 'Farmer First Revisited', December 2007. Available at: http://www.future-agricultures.org/farmerfirst/files/T3c_Hagmann.pdf (accessed July 2008).

Hagmann, J., C. Almekinders, C. Bukenya, F. Guevara, A. Halemichael, P. Isubikalu, G. Kamau, G. Kamanga, P. Kibwika, B. Limnarankul, F. Matiri, T. Mutimukuru, H. Ngwenya, C. Opondo, L. Zhang and U. Breitschuh, 2003. Developing 'soft skills' in higher education. PLA Notes 48 (Dec 2003): 21-25.

ICRA/NATURA, 2003. Proceedings of ICRA-NATURA workshop on Mobilising partnerships for capacity building in integrated agricultural research for development (IAR4D). Wageningen, the Netherlands.

Kibwika, P., 2006. Learning to make change. Developing innovation competence for recreating the African university of the 21st century. Wageningen Academic Publishers, the Netherlands.

Levin, M. and A. Martin (2007). The praxis of educating action researchers. The possibilities and obstacles in higher education. Action Research 5: 219-229.

Mutimukuru, T., R. Nyirenda and F. Matose, 2005. Learning amongst ourselves. Adaptive Forest Management through social learning in Zimbabwe. In: Colfer, C.J.P. (ed.). The equitable forest: diversity, community and resource management. Resources for the Future and CIFOR, Washington DC., USA, pp. 186-204.

Patel, K.B., M. Maina, J. Hagmann and P.L. Woomer, 2001. Curriculum development and transformation in rural development and natural resource management: a strategy workshop conducted at the Rockefeller Foundation's Bellagio Center in Italy, November 12-16, 2001. Rockefeller Foundation Nairobi.

Pretty, J., I. Guijt, J. Thompson and I. Scoones, 1995. Participatory learning & action. A trainer's guide. IIED, London. 267 pp.

Sanginga, P.C., A. Waters-Bayer, S. Kaaria, J. Njuki and C. Wettasinha, 2008. Innovation Africa enriching farmers' livelihoods. Earthscan, London. 384 pp.

Starks, A., M. Jones and R. von Kaufmann, 2005. The sub-Saharan Africa challenge program: an experiment in mainstreaming institutional learning and change. ILAC Brief 9. Available at: http://www.cgiar-ilac.org/downloads/Briefs/Brief9Proof2.pdf (accessed July 2008).

6. Taking a stance

The action researcher as an independent actor

Rudi Roose and Maria de Bie

6.1 Introduction

The ambition of action research is often the ambition of emancipatory research, that is the research should contribute towards changing the situation under investigation, and this change should involve an improvement. It is often claimed that emancipatory research should be viewed as democratic research: it does not involve the implementation of changes desired by the researcher in social reality, but it involves a societal change induced by a democratic research process in which not only the researcher but also policymakers and clients take an active part. The present article analyses the scope and the concrete implementation of the 'emancipatory' action research we conducted. We put 'emancipatory' between quotation marks since the term can have several meanings, as we will see below. The first section gives a brief description of the action research we conducted. Next, we will review the features we believe to be essential to emancipatory action research: the focus on the growing problem content of the situation, and the starting point of a subject-subject relationship between researchers and other actors participating in the research. We will discuss these features basing ourselves on the findings of the action research we conducted. This analysis will show that emancipatory research is subject to several preconditions, which are hard to achieve in the practice of action research, however. One of the conclusions is that it is the action researcher's task to help to materialise these conditions. This presupposes a modest attitude linked to a strong involvement with the problem situation that is the subject of the research.

6.2 The action research 'Youth Care Waasland'

From April 2001 to the end of June 2003, we conducted the action research 'Youth Care Waasland'. The research was induced by the wishes of several youth care organisations in the Flemish region called Waasland. They asked us to support the reorganisation of youth care which was promised by the government at the time. This reorganisation aimed at a better mutual organisational rapport between youth care organisations in order to achieve a user-led youth care organised at the intersectoral level; the overall aim was to make youth care more accessible.

The research involved 35 youth care organisations. The actors from the region concerned expressly expected the reform of youth care to benefit the children and the parents in the region. At the same time, they expected a bottom-up approach which took account of the existing cooperative culture in the region and which fitted in with concrete practice. The desire to gear the research to practice and to the demands and objectives of the parties involved necessitated a participative approach to the research. In consultation with the actors, namely representatives of youth care services in the region, it was decided to launch an *action research*. This choice satisfied the demand for support for the envisaged process of change and contextualised knowledge (Tromp, 2004). The starting point is that knowledge cannot be acquired unless you try to change something: *if you want to know how things really are, just try to change them* (Lewin, in Schruijer, 2005). Only by initiating a process of change can we discover any resistances, identities, possibilities and impossibilities (Schruijer, 2005).

6.3 Action research: the growing problem content

Noffke argues that the history of action research is characterised by the tension between 'democracy and social engineering, which continues to be worked out in changing configurations in the practice of action research' (Noffke, in Altrichter and Gstettner, 1997: 61). In the social-engineering approach, the social position of the research is not fundamentally questioned, in contrast to emancipatory research. Our research was conceived as emancipatory research. A possible criticism of this description of our research, and of action research presented as emancipatory in general, is that action research emphasises the interpretation of the current situation instead of actually seeking to change this situation (Romme, 2004). Romme argues that action research aimed at emancipation requires more *design research*, which may contribute to the achievement of emancipatory objectives. This proposition implies a conception of action research as a *process of innovation* aimed at a specific *Bildungsideal* (Masschelein, 1991), namely that of a *state of emancipation* to be reached. We think it thus again becomes a type of social engineering, involving the implementation by the researcher of a democratic model, not a change that is jointly achieved and supported through a democratic research process. A democratic research process by definition implies an open research process to which various actors, and not just the researcher, make their own contribution.

Our action research started from an approach to emancipation in line with the pedagogical thinking of Paolo Freire. Freire was a major source of inspiration to us, since he emphasised cooperation rather than the expectation that people *should* emancipate themselves through this cooperation (Van Hove, 1999). 'Paulo Freire's central message is that one can know only to the extent that one 'problematises' the natural, cultural and historical reality in which s/he is immersed. Problematising is the antithesis of the technocrats' 'problem-solving' stance. In the latter approach, an expert takes some distance from reality, analyses it into component parts, devises means for resolving difficulties in the most efficient way, and then dictates a strategy or policy' (in Freire, 2005: ix). The intended change in our action research therefore involved a growing understanding of the situation (Bouverne-De Bie, 1988). This meant that the adequacy of the *starting points* of the reorganisation, and not just the organisation of youth care in itself, became central to the research. In a prior consultation with the representatives of the youth care organisations, the starting point of the action research was defined. The starting point was that a reorganisation of youth care does not by definition generate an added value for the children and the parents. Cooperation between care organisations may also result in clients receiving less adequate support options. For instance, when network development leads to a one-sided reinforcement of the position of social workers, it can have the effect that the views of those requesting help are ignored. The initial research question was 'how can youth care be reorganised so that this reorganisation also offers an added value to the clients?' 'Added value' was understood to mean youth care that meets the criteria of availability, accessibility, affordability, usefulness and comprehensibility. *Availability* refers to the existence of a supply and to the fact that social services can be called upon for matters that do not relate directly to the assessed problem. *Accessibility* refers to the (lack of) thresholds when care is needed, for instance an inadequate knowledge of the supply. *Affordability* refers to financial and other costs that the client may encounter, for instance giving up one's privacy or negative social and psychological consequences of an intervention. *Usefulness* refers to the extent to which the client experiences the care as supportive: is the help attuned to the demand, the skills and the language of the clients? *Comprehensibility* refers to the extent to which clients are aware of the reasons for the intervention and the way in which the problem should be approached (Roose and De Bie, 2003).

These criteria were considered essential to viewing youth care as a type of service provision which realises potential conditions in view of developing a decent existence. The objective of a decent existence is enshrined in the International Convention on the Rights of the Child, which was the frame of reference used by both youth care and the research.

6.4 The subject-subject relationship in action research

Action research is conducted in cooperation with the parties involved. The literature describes this relationship as a subject-subject relationship. This subject-subject relationship between the researchers and the researched subjects requires that special attention should be paid to the cooperation agreements. The proposition that action research involves a subject-subject relationship between the researchers and the researched subjects fails to answer the question as to who these researched subjects are and what kind of subjects they are expected to be. Traditionally, research distinguishes between four parties (Wadsworth, 1998): (1) the researcher; (2) the researched subjects; (3) those who should benefit from the research, the 'critical reference group' – in our case the children and their parents; and (4) others who stand to benefit from the research, for instance policymakers.

The organisations participating in the action research did not wish to reorganise youth care without consulting the clients, and from the outset of the research they raised the question as to how children and parents could be actively involved in the reorganisation of youth care. This active participation of children and parents in youth care and its development is deemed important: client participation is believed to have an empowering effect, to improve the balance of power between social workers and clients, and to increase the freedom of children and parents. 'Freedom' in this context refers to the freedom to make decisions, to take one's own life into one's hands (Holland and Scourfield, 2004). However, the emphasis on the active participation of children and parents may also have adverse effects (Hayward *et al.*, 2004), since the participatory discourse (the objectives and values) and the participatory techniques (the way in which participation is materialised) evoke a highly *normative subject image* of an autonomous and competent individual (Masschelein and Quaghebeur, 2005). If clients fail to participate, this is construed not so much as their right to offer resistance in the power relationship with their youth carers – *the power to disbelieve* (Anderson, in Pease, 2002) – but as part of their problem: non-active clients are seen as problem clients. The issue of client participation should therefore be approached cautiously, given the possibly counterproductive effects of a participatory approach and the risk of instrumentalising children and parents when they are involved from within the youth carers' frame of reference (Beresford, 2000). That is why it was decided, in consultation between the researchers and the youth care services, to set up our action research from a different perspective. We viewed participation as a *starting point* of care: a participatory care policy is a policy in which children and parents are acknowledged and recognised. This implies that care, in all its aspects, should be developed from the perspective of the children and the parents. The question is how this perspective is to be introduced. This appeared to be much more complex than it would seem at first sight (cf. infra).

When participation is taken as the starting point, the focus of the research shifts from the *actions of the children and the parents* to the *actions of the youth carers*. This implies that the object of the action research is the *care policy* and that the subject-subject relationship between researchers and researched subjects should be a subject-subject relationship between the researchers and the youth carers involved.

The subject-subject relationship (or the striving for such a relationship) was given organisational shape in a steering group consisting of the researchers and staff members of the participating youth care organisations. This steering group was considered the main lever for achieving a process of change at the policy level. At the same time, an executive committee and a partnership were set up. The executive committee consisted of a researcher and several other members of the steering group, whilst the partnership involved all the organisations that participated or were interested, as well as their staff and the subsidising authorities. During the first meeting of the steering group, a project leader was appointed for whom a fulltime function could be created, thanks to the financial support of the local authorities. The project leader was one of the initiators, a director of a special youth care service who had many years' experience and was highly regarded in the region.

Methodologically speaking, the research was conceived as a movement towards the problem field, in which we increased the level of concreteness of the youth care discussion by examining concrete practice (Roose and De Bie, 2003). The first phase involved the screening of what the services within the region expected of the results of the reorganisation and a survey of these expectations in terms of what they could mean to the clients. In a first stage, the client perspective was introduced indirectly via a review of the literature. On the basis of this screening, several actions points were formulated and the findings were brought together in a number of thematic working groups, for instance on how to deal with client information, on participatory care, and on the transparency of the offer to the clients. In a second stage, the work of these thematic working groups was deepened by means of concrete case reviews: initially the discussion of past cases, but subsequently of current cases as well, including the direct input (through interviews) of the clients' perspective.

6.5 Findings from the action research

From the reorganisation of care to the study of care practices

During the action research we quickly found that it was not easy to introduce the client's perspective. The initial question started from the reorganisation of care and entailed a rather abstract discussion. In order to obtain a view of the clients' perspective, it was necessary to investigate concrete care practice. And this implied a *movement towards the problem field* (cf. supra). This was done on the basis of the interim research results.

As suggested above, the methodological concept involved an increasingly concrete movement from the research towards the problem field. This concretisation made the client increasingly 'visible' in the research. In the first stage of the action research, this was done indirectly, as described above, by drafting and discussing a review of the literature on clients' perspective on care. In a later stage, this was done directly by interviewing clients and discussing concrete cases. In each of these stages, the same questions were addressed. However, this concretisation implied that the questions had to be discussed in a decreasingly abstract way and were to be linked more and more to concrete care practice.

In other words, the researcher also rephrased the original research question (the reorganisation of care) into a reflection on and an investigation of the carers' own practice from the client's point of view. In the steering group meetings, the reflection on the carer's own care practice was always related to the reorganisation, and notably to the starting point that a reorganisation should generate an added value for the client. This rephrasing met with resistance from the organisations concerned: for instance, the request for supplying cases yielded a very poor response. During the discussion of this response, the actors involved claimed that they had no scope for answering our question. Our questions referred to the possibility and the willingness *to assume a vulnerable position and to fundamentally question one's own starting points* and discuss them with other organisations. They said there was no scope for this, since it would cause too much uncertainty or resistance. Care appears to be characterised by a conception of expertise which emphasises the capacity to resolve and control problems which is incompatible with a vulnerable position.

In the light of these findings, it was decided, jointly with the organisations involved, to cancel the action research in its current form, given the limited scope for change that appeared to exist. Finally, a colloquium was organised where the results of the action research were presented to the public and where a number of topics that had been covered by the research were further elaborated in workshops.

From research as helpful to action and as a means for action, to action as a means for research

How should we evaluate our action research? One could argue that it has failed, since it has not produced the desired solution to the actors' problem, namely a reorganisation of youth care. However, from a research point of view, the research produced highly interesting results. The resistance to the proposed changes gave us – as indeed Lewin suggests – an insight into why change is so difficult to accept. It provided a better understanding of the idea of expertise as helpful to youth care, which focuses on a knowing and predictable care. Carers are deemed to be experts at solving the problems they are confronted with. The emphasis is on controlling problems rather than understanding them. This is at odds with the perspective of the clients, who seek flexibility and easy accessibility. It is also at odds with the finding that a lot of problems cannot be controlled by youth care, since their cause, for instance poverty, is often beyond the reach of youth care. Taking account of the perspective of children and parents consistently would therefore make it possible to break through the status quo, with carers assuming a more modest attitude and questioning their own actions and ideas. Carers suggested that this was too hard to achieve in the present social and youth care context.

Throughout the research, *parallels* appeared between the relationships between researchers and carers and carers and clients. The researcher should display the same modesty vis-à-vis 'co-researchers' as the carer vis-à-vis the client. Researchers should also mitigate their emancipatory or liberating pretensions (Healy, 2001). However, this modesty should not cause a lack of commitment or a form of *creaming*, in which only research that is expected to be successful is engaged in.

A question that could be raised is whether the conditions for launching our action research were present. Indeed, we were facing a highly complex research situation with regard to both the research questions and the scope of the research field. Zeelen (in Boog, 2002) for example points to the need for an interests analysis prior to involving the researched subjects in a research process, his conclusion being that it is sometimes better not to launch a research project at all. However, we believe that these interests can only be expressed *in the course of the research*, since otherwise we would have to specify beforehand that the research is to induce concrete change.

This relationship between research and action is an *interchange* in which different aspects can be distinguished (Bouverne-De Bie and Verhellen, 1995): (a) *The research is a means for action*, since it questions existing approaches vis-à-vis a given problem. This is in line with the formative function of the research. In our action research this was, for example, achieved by organising training moments which offered an occasion for reflecting on the thresholds impeding quality actions; (b) *The research is helpful to the action,* since it supplies cognitive data that may support the action. This is the counselling function of the research. In our action research, we could refer to the literature review on *Clients on Formal Care* (Roose, 2001); (c) *The action is a means for research.* The steps taken in the action research gave us a better understanding of the problem of youth care. Emancipatory action research requires these three elements to be balanced, and this balance in its turn requires the researcher to be able to adopt an *independent attitude.*

The different expectations and positions of researchers and practitioners

It goes without saying that the researcher is largely dependent on the cooperation with the other actors in the research, since action research involves a subject-subject relationship. This is not a one-to-one relationship between one researcher and one researched subject, but a relationship between several researchers and several researched subjects. Our action research made a distinction between all these *subjects*: (1) two university researchers; (2) the organisations concerned – staff members and the *frontline workers*; and (3) the project leader.

In our action research, the researched subjects were called *co-researchers* (Lammerts, 1998). This implies a different setup than the traditional research approach, in which the researcher says something *about* practice. Action research consequently views the researched subject not only as a major actor in knowledge production, but also as an actor in the *reflection about this knowledge*. It is research *with* and not just *about* practice. At the same time, we should remember that the researcher and the co-researchers take different positions in practice and in knowledge production, for instance in reporting, where it is the researcher who drafts the reports.

In action research, the position of the researchers and co-researchers can be situated on the axis between the *insider and outsider position* (Van der Kamp, 1996). Van der Kamp defines the researcher as an 'engaged and competent outsider' (Van der Kamp, 1996: 122). On the one hand, this outsider role can be problematical and cause resistance, for instance if the proposed changes are not supported by the co-researchers. On the other hand, an insider position can lead to 'a canonisation of the research subject' (*ibid.*: 124), causing any objectivity or critical distance to disappear.

A possible solution to this insider/outsider tension could be to split the research task through a *double act relationship* (Titchen and Binnie, 1993), in which the roles can be distributed if there are several researchers: one researcher could be emphatically present in the field (insider), whilst the other researcher could take a more distant, coordinating role (outsider) (cf. Aertsen and Decraemer, 1993).

Instead of a double act relationship, our action research relied on a *triple act relationship*, in which the researchers and the project leader took different positions. The project leader's task was to monitor the formal aspects of the project (contacts with the actors, organisation of day-to-day business, finances, contact point for the steering group, etc.), and to carry out research tasks (conducting screenings, setting up trainings, etc.). His main task consisted in effecting change from within and creating and monitoring a basis for the research. Action research not only involves a dialogue: in practice, it is often a *multilogue* that takes place (Mölders, 2001). Boog (2002) argues that action research is a living social network. The project leader's role is to explore this network and the communication flows about the research, and to respond appropriately. The importance of the project leader's position became even clearer when another project leader replaced him. The new project leader worked halftime instead of fulltime and held a different, less authoritative position in the region.

The *project leader* was positioned in between an insider and an outsider position. The organisations concerned probably viewed him as an insider mainly, while the researchers sometimes also placed him in the outsider position.

Both researchers took different positions. The *first researcher* worked on the tension of an insider/outsider position. He performed various tasks as a *committed outsider*, based on the components referred to above, notably a counselling function, a training function and a research function. Apart from this outsider function, the researcher was closely involved with the concrete practice of the project and he approached the *insider position* closely. The fact that he had been active in youth care himself for many years and knew several of the actors from his own professional practice, generated a tension – which was also caused by the close contact with the practitioners – to become involved in the day-to-day concerns of practice and the need to address these concerns.

This tension was countered by the fact that the *second researcher,* who was the supervisor of the research, took an outsider position as chairman of the steering group. She remained at a greater distance from concrete practice and thus succeeded in monitoring the line of the research by positioning herself as an outsider. The decision to entrust the chairmanship to a researcher was motivated by the need to monitor the frame of reference of the research and by the assumption that this requires some distance. The demand of the researchers to be given the chairmanship of the steering group was felt by the practitioners as a relinquishment of the project. The researchers indicated that it was essential to define the different positions clearly and that the chairmanship of the steering group – an important research tool –

should go to the researchers. At that moment, the difficulty of achieving a subject-subject relationship became evident.

The positions taken by the different actors in the research are related to the tension between *involvement* and *distance*: apart from involvement, there should also be a certain distance vis-à-vis the problems presented by practitioners (Elias, 1982). Elias argues that the development of social-scientific knowledge is characterised by a lack of distance and a too profound emotional involvement in the problems. This excessively profound involvement makes it impossible to step back from the problem, viewing the problem in a broader context and seeking sustainable instead of instant solutions. This lack of distance thus produces unsatisfactory answers to the problems raised, and this in its turn reinforces the emotionality vis-à-vis the problem and thus the pressure to control the problem. Although action research emphasises the recognition of and attachment to involvement, it also deems distance to be important, since this makes it possible to ask more adequate questions rather than give answers immediately.

The idea of engaging in a subject-subject relationship may be wishful thinking (Coenen, 2001). This is related amongst other things to the possibly different expectations with regard to the research. As said, action research has a *dual aim*: raising the competences of practitioners and theory building (Boog *et al.*, 1996). It is obvious that it will be mostly the practitioners who will pursue the first aim, whilst the researcher's interest will focus mainly on the second aim. Whilst practitioners will want to resolve the problem, researchers will essentially seek a better understanding of the problem. These two approaches need not be at odds with each other, but they can generate tension, because the research runs counter to the 'illusion of taking action' (Senge, in Mensink, 2005: 30): seeking to solve any problem you are confronted with directly instead of making an in-depth problem analysis first by constantly referring to the starting points underlying the problem situation. This becomes even more prominent when we take Lewin's proposition referred to above seriously – *if you want to know how things really are, just try to change them* – this approach views action research as a type of *double-loop learning* (Argyris, 1996), in which the basic assumptions of youth care are questioned time and again, in contrast to single-loop learning, in which the problems are tackled without questioning the basic premises.[9]

The researcher views the practitioner as a subject who is to be taken seriously and who is an expert in his domain; however, this also implies the expectation that the practitioner as a subject will think along and reflect on his own practice. This is not evident: practitioners are caught in a certain discourse on what an expert is or should be (Margolin, 1997), and the expectation which is inherent to the setup of the action research, namely that carers are *reflective practitioners*, is not self-evident. The action research suggests that the approach

[9] Argyris and Schön also distinguish *deuteron learning*, in which organisations themselves learn how to control their learning processes, in other words, *learning to learn* (Mensink, 2005).

to carers as reflective practitioners involves a learning process to which action research can make a contribution. At the same time, the research shows that this contribution cannot be presupposed but has to take shape during the research itself.

6.6 The context of the research

The context in which the research relationship is established has major consequences for the relationship between the researcher and the researched subject, as well as for the possible positions they can take. Importantly, our action research was no *contract research*: it was conducted at the request of the practitioners but was not funded by them: it was funded by the University. In this way, an independent position vis-à-vis the practitioners could be taken and the research objectives could be steered by the researchers. The instrumentalisation of the actors in action research is often caused by the fact that the research is part of a specific policy mission which lays down the objectives to be achieved beforehand (Keune, 1998). In this case, the researchers make use of the practitioners. Conversely, researchers can also be asked by practitioners *to solve* problems arising in practice[10], in which case they can also be *used* by practitioners.

Our action research was the result of a request by practitioners which the researchers could fulfil in the framework of a PhD project. This gave practitioners the prospect of being able to tackle the reorganisation of youth care in a more carefully thought-out way. According to the researchers, the request by practitioners offered an opportunity to gain access to the field of practice. Other elements can also play a role, often implicitly (Zeelen, 1994): for instance, for the region it was important that it could profile itself as a region which tackled the reform process scientifically.

The *formal structuring* of the dialogue between the researchers and the researched subjects was kept to a minimum. The organisations involved had commissioned the research, and since the research took place in the framework of a PhD project, the researchers' wages were paid by the University; the research costs borne by the region remained limited to the costs for transportation and photocopies. As a result, both the researchers and the researched subjects were given the scope to address *fundamental questions*: the scope for conducting a fundamental dialogue widened, and the researchers and researched subjects had the option to step out of the research at any time if a common ground for further collaboration was no longer present.

[10] For instance, practitioners can request an evaluation research of their own operation which is conducted in the framework of an application for subsidies. The scope for assessing practice critically thus becomes very small.

6.7 Conclusion

Emancipatory action research displays high ambitions in the literature. Boog for instance says with regard to the emancipatory objectives of action research: 'Action research is designed to improve the researched subjects' capacities to solve problems, develop skills (including professional skills), increase their changes of self-determination, and to have more influence on the functioning and decision-making processes of organisations and institutions from the context in which they act' (Boog, 2003: 426). Reason (2001) argues that action research has a dual objective: to acquire knowledge and actions which can be applied directly by the actors involved, and to enable the actors to gain a better insight into the matter and to empower them; to use their own knowledge to initiate a learning process towards becoming a 'reflective practitioner'.

The concept of empowerment starts from an assumption of power as something which is distributed unequally and which should be distributed more equitably (Dahlberg and Moss, 2005). Reason and Bradbury for instance argue that the pedagogy of the oppressed (Freire) should be matched by a *pedagogy of the privileged*: 'inquiry processes which engage those in positions of power and those who are simply members of privileged groups – based on gender, class, profession, or nation. We need to learn more about how to exercise power and position legitimately in the service of participative relationships, to find ways in which politicians, professionals, managers can exercise power in transforming ways, power with others rather than power over others' (Reason and Bradbury, 2001: 10 referring to Torbert). Hence, a dichotomy is stated between the powerful and the powerless. This dichotomy might ignore the fact that, as our research shows, carers might also be powerless to a certain extent.

This conception of power involves other dichotomies besides the distinction between power/powerlessness and privileged/poor: carer/client, technical knowledge/experience-based knowledge, but also research/researched subject (Healy, 2000). These dichotomies might lead to the assumption that care can empower the client and that research can empower the carer. The emancipatory ambitions should be tuned down, since power is not a commodity that can simply be transported from the researcher to the researched subject, or from the carer to the client. More likely, the carer, the client and the researcher are all caught in a *complex network of power relations* (Foucault, 1988) in which they have to move and in which every movement generates new power relations.

Rather than being a large-scale empowering project, action research is nothing more than a type of *minor politics* (Dahlberg and Moss, 2005) in which reality as we experience it is questioned and the roles we play are reviewed critically. The researcher's ambitions should therefore be approached with due modesty. Not because any resistance is to be overcome, for instance to the image of the researcher as an *expert* who presents the solution, but because of the 'slow dynamic of the (re)production of social reality' (Coenen, 1987: 165): social changes are slow to take effect.

The idea of emancipatory research stresses the dialogue between researchers and researched subjects. This *dialogical model* is opposed to a *normative approach* to emancipation, in which the objective of emancipation is defined beforehand as a result to be achieved (Leonard, quoted in Pease, 2002). The dialogical and normative approaches are not always easy to distinguish in practice. There is a *field of tension* in which action research can position itself in varying ways during the research (Keune, 1998). A lot of action research for instance makes use of the theory building about the *learning organisation* (Argyris, 1996), in which the focus is on effective action. The pragmatic approach uses this theory building self-referentially: there are no external standard and frames of reference for effectiveness (Veendrick and Zeelen, 1997; Mensink, 2005), or else these frames of reference are imposed from the top down. This implies an instrumental relationship between the researchers and the researched subjects: although they do enter into a dialogical relationship, they pursue *external* objectives. The dialogue in action research thus becomes a kind of window dressing in which participatory techniques are applied as a strategy for achieving one's own objectives (Van Diesen, 1998), either by the researcher or by the researched subjects. The *emancipatory power* of action research should therefore be looked at from different angles, as should the idea that it is obvious that the relationship between researcher and research subject is a subject-subject relationship or that the researched subjects are really co-researchers. We believe that such an instrumental relationship runs counter to our approach to emancipatory research.

The experiences described above show that it is important to reflect on one's position as a researcher and to realise that emancipatory research requires a modest attitude, but also a stubborn attitude which does not yield to the daily illusions, but in which existing tension relationships – between emancipatory and pragmatic research, between the insider and outsider positions – are kept open. This can cause resistance and may even mean the end of the research. Paradoxically, ending the research can be viewed as an element of its participatory and open nature: the actors involved are entitled to offer resistance and they should not necessarily follow the proposals and the problem analysis presented by the researchers. It is the responsibility of the researchers to keep the dialogue open, and it is the responsibility of youth carers to decide what they will do with the research results.

References

Aertsen, I. and K. Decraemer, 1993. Geestelijke gezondheidszorg en slachtoffers van crimineel geweld. Eindrapport over een actiegericht onderzoek in samenwerking met vijf centra voor geestelijke gezondheidszorg. Catholic University Leuven, Leuven.

Altrichter, H. and P. Gstettner, 1997. Action research: A closed chapter in the history of German social science? In: R. McTaggart (ed.), Participatory action research. International context and consequences. State University of New York Press, Albany, pp. 45-78.

Argyris, C., 1996. Leren in en door organisaties. Het hanteerbaar maken van kennis. Scriptum Books, Schiedam.

Beresford, P., 2000. Service users' knowledges and social work theory: Conflict or collaboration. British Journal of Social Work 30: 489-503.
Boog, B., 2002. Handelingsonderzoek: een update. Sociale Interventie 11 (4): 27-40.
Boog, B., 2003. The emancipatory character of action research, its history and the present state of the art. Journal of Community & Applied Social Psychology 13: 426-438.
Boog, B., H. Coenen, L. Keune and R. Lammerts (eds.), 1996. Theory and practice of action research. With special reference tot the Netherlands. Tilburg University Press, Tilburg.
Bouverne-De Bie, M., 1988. De zorg voor jeugdwelzijn als overheidsopdracht. Een studie naar de aard van de geïnstitutionaliseerde jeugdwelzijnsinterventie. Proefschrift ter verkrijging van de graad van doctor in de staats - en bestuurswetenschappen. Rijksuniversiteit Gent, Gent.
Bouverne-De Bie, M. and E. Verhellen, 1995. Een beweging naar het probleemveld toe: actieonderzoek en straathoekwerk als 'doelgroepgerichte' interventies. Panopticon 16: 176-189.
Coenen, H., 1987. Handelingsonderzoek als exemplarisch leren. Konstapel, Groningen.
Coenen, H., 2001. Handelingsonderzoek: de verhouding tussen onderzoeker en onderzochte. Tijdschrift voor Arbeid en Participatie 23 (1): 63-74.
Dahlberg, G. and P. Moss, 2005. Ethics and politics in early childhood education. Routledge Falmer, Oxfordshire.
Diesen, van, A., 1998. Keeping hold of the stick and handing over the carrot: dilemmas arising when development agencies use PRA. In: B. Boog, H. Coenen, L. Keune and R. Lammerts (eds), The complexity of relationships in action research. Tilburg University Press, Tilburg, pp. 37-50.
Elias, N., 1982. Problemen van betrokkenheid en distantie. Meulenhoff, Amsterdam.
Foucault, M., 1988. On power. In: Kritzman, L.D. (ed.) Michel Foucault. Politics, philosophy, culture. Interviews and other writings 1977-1984. Routledge, New York, pp. 96-109.
Freire, P., 2005. Education for critical consciousness. Continuum, London, New York.
Hayward, C., L. Simpson and L. Wood, 2004. Still left out in the cold: Problematising participatory research and development. Sociologia Ruralis 44: 95-108.
Healy, K., 2000. Social Work. Contemporary perspectives on change. Sage, London, Thousand Oaks, New Dehli.
Healy, K., 2001. Participatory action research and social work. A critical appraisal. International Social Work 44: 93-105.
Holland, S. and J. Scourfield, 2004. Liberty and respect in child protection. British Journal of Social Work 34: 21-36.
Hove, van, G., 1999. Coöperatief onderzoek met personen met een verstandelijke handicap. Een illustratie van de emancipatorische orthopedagogiek. Tijdschrift voor Orthopedagogiek, Kinderpsychiatrie en Klinische Kinderpsychologie 24: 24-33.
Kamp, van der, M., 1996. Research on social intervention: problems and perspectives. In: B. Boog, H. Coenen, L. Keune and R. LammertS (eds.), Theory and practice of action research. With special reference to the Netherlands. Tilburg University Press, Tilburg, pp. 119-128.
Keune, L., 1998. The strategic role of adequate relationships in action research-Final remarks. In: B. Boog, H. Coenen, L. Keune and R. LammertS (eds.), The complexity of relationships in action research. Tilburg University Press, Tilburg, pp. 195-201.

Lammerts, R., 1998. Activating research. In: B. Boog, H. Coenen, L. Keune and R. Lammerts (eds.), The complexity of relationships in action research. Tilburg University Press, Tilburg, pp. 81-98.

Margolin, L., 1997. Under the cover of kindness. The invention of social work. University Press of Virginia, Charlottesville.

Masschelein, J., 1991. Zijn er nog overlevingskansen voor het 'Bildungsideal'? Pleidooi voor vorming die niet 'werkt' en niets 'verbetert'. Vorming 7: 39-52.

Masschelein, J. and K. Quaghebeur, 2005. Participation for better or for worse. Journal of Philosophy of Education 39: 51-65.

Mensink, H., 2005. Chris Argyris: de lerende organisatie in de bestuurskamer. In: B. Boog, M. Slagter, I. Jacobs-Moonen and R. Meijering (red.), Focus op action research. De professional als handelingsonderzoeker. Koninklijke Van Gorcum, Assen, pp. 18-36.

Mölders, H., 2001. The process of developing an improved understanding: Experiences with a communication project on mental suffering. In: B. Boog, H. Coenen and L. Keune (eds.), Action research: Empowerment and reflection. Dutch University Press, Oisterwijk, pp. 147-174.

Pease, B., 2002. Rethinking empowerment: a postmodern reappraisal for emancipatory practice. Sociale Interventie 11 (3): 29-37.

Reason, P., 2001. Learning and change through action research. In: J. Henry (ed.), Creative management. Sage, London, pp. 182-194.

Reason, P. and H. Bradbury, 2001. Inquiry and participation in search of a world worthy of human aspiration. In: P. Reason and H. Bradbury (eds.), Handbook of action research. Participative inquiry and practice. Sage Publications, London, Thousand Oaks, New Dehli, pp. 1-14.

Romme, A.G.L., 2004. Action research, emancipation and design thinking. Journal of Community & Applied Social Psychology 14 (6): 495-499.

Roose, R., 2001. Cliënten over formele hulpverlening. Onuitgegeven rapport. Universiteit Gent, Gent.

Roose, R. and M. De Bie, 2003. From participative research to participative practice - a study in youth care. Journal of Community & Applied Social Psychology 13: 475-485.

Schruijer, S., 2005. Op weg naar een waardevolle wetenschap. In: B. Boog, M. Slagter, I. Jacobs-Moonen and R. Meijering (red.), Focus op action research. De professional als handelingsonderzoeker. Koninklijke Van Gorcum, Assen, pp. 7-17.

Titchen, A. and A. Binnie, 1993. Research partnerships: Collaborative action research in nursing. Journal of Advanced Nursing 18: 858-865.

Tromp, C., 2004. Breedbeeld wetenschap Een kritisch-reflexief onderzoeksmodel gebaseerd op een breed rationaliteitsbegrip. Van Arkel, Utrecht.

Veendrick, L. and J. Zeelen, 1997. Sociaal leren in de context van sociale uitsluiting. Pedagogisch Tijdschrift 22: 39-52.

Wadsworth, Y., 1998. What is participatory action research? Available at: www.scu.edu.au. Accessed 23 April 2003.

Zeelen, J.J.M., 1994. Handelingsonderzoek en programma-evaluatie. Sociale Interventie 3 (1): 21-27.

7. Searching for dialogue with the Zapatistas

Dilemmas of collaborative research in the ethnography of conflict

Gemma van der Haar

7.1 Introduction

Over the past decade a number of fascinating proposals have been developed to make anthropology more responsible and responsive to the populations studied. This has happened under labels such as 'collaborative' or 'activist' anthropology (Hale, 2001), 'militant anthropology' (Scheper-Hughes, 1995) or the 'anthropology of witnessing' (Stephen, 2002: 31). These views take a step back from the image of the anthropologist as the 'neutral observer' whose main task it is to document social processes, to posit the anthropologist instead as a person who commits him or herself to disadvantaged groups, to respond to suffering and to denounce injustice and abuse.

In this chapter I engage with these proposals and the promises, challenges and potential problems they entail in the practice of field research. In particular, I explore the consequences, both positive and possibly problematic, of such an approach in contexts of social and political conflict. I do this not from a safe abstract vantage point but by directly questioning my own practice, and especially the way I set up and conducted my field research into conflict and municipal government in Chiapas (Southern Mexico). My research was not designed as 'action research' or 'activist anthropology', but as the project developed I became increasingly interested in the activist and collaborative approaches in anthropology. The alternative anthropologies discussed have not been a distant scholarly debate to me, but on the contrary, during my work I interacted with many anthropologists who embraced activist and collaborative methodologies.

On the basis of the 'activist' and 'collaborative' propositions, I started to look critically at my own research practice, particularly the way in which I approached the main conflicting parties in the local setting of my fieldwork. My interest in collaborative approaches to anthropology is related, on the one hand, to the difficulties I encountered in doing fieldwork and the continuous doubts as to whether I had made the right choices in connecting with local groups. On the other hand, it is motivated by my conviction that in many ways, collaborative research is the anthropology of the future. It addresses a number of key issues in contemporary ethnography, including (1) the need to be accountable to the subjects studied not just as a moral imperative but in order to be able to gain access to the research population in the first place; (2) the need to reflect on broader agendas of change with which the research will, intentionally or unintentionally, interact; and (3) the need to reflect on the moral and ethical responsibilities of anthropologists. At the same time, I have some doubts about some aspects of collaborative approaches, especially in the context of political conflict. I see potential problems with the possible silencing of counter-narratives or marginal voices within and around the organisation, arising from the close and exclusive association of the researcher with one particular organisation.

The main aim of this contribution is to share ideas and doubts. This text does not advocate any approach in particular or explain how to 'do' collaborative work. Rather, it is meant as an input to help the reader decide what kind of ethnographer he or she wants to be, whether a collaborative approach would be feasible and productive for the kind of questions he or she plans to address and the context in which this will be done, and to sensitise him or her to some of the issues that might be encountered along the chosen path. The account presented here thus hopes to stimulate reflection on how to give shape to ethnographic research. It is inspired by the reflexive nature of much current anthropology (whether explicitly collaborative or not) that includes the experiences and concerns of the researcher in the ethnographic narrative. I think such reflexive narratives are important ways of exploring the often confusing and multi-layered social processes which anthropologists try to understand and unavoidably become part of and therefore have added value for the analysis.

In the remainder of this chapter I first discuss some of the main claims and proposals of collaborative anthropology. I then introduce my own research in Chiapas. Next, I reflect upon a number of specific issues on which collaborative anthropology has something to say: negotiating access to the field, dealing with multiple perspectives, and the commitment of the researcher. I conclude that collaborative research is likely to afford a more profound analysis as it lends the interaction between researcher and researched more urgency, but that in addition we should think out other ways to develop a responsive and responsible anthropology.

7.2 Activist and collaborative anthropologies[11]

Collaborative, activist or other politically committed anthropologies do not form a homogeneous or 'finalised' school of thinking.[12] It is work-in-progress which engenders a rich and often passionate debate. Different emphases may be encountered amongst authors.[13] Lynn Stephen and Nancy Scheper-Hughes, for example, stress the moral responsibility of the anthropologist not to remain indifferent to human suffering but to denounce and react.

[11] This section draws on a brief unpublished paper which I wrote in 2005 (Van der Haar, 2005c).

[12] There is also no suggestion that these ideas are entirely new. Activist and collaborative anthropology build on earlier ideas but elaborate on them and have given them a renewed significance in relation to the present conditions.

[13] I make no claims to an exhaustive discussion. My starting point in this discussion has been the ideas of Charles Hale and the activist anthropology promoted in Austin; other authors consulted are Stephen, Simonelli and Earle, and Scheper-Hughes. Many others could be added, especially as more and more PhDs are now being completed based on these ideas.

Charles Hale (2001) and the programme in activist anthropology at the University of Texas at Austin emphasise the commitment to progressive social and political transformation and the collaboration throughout the research process with social actors. The 'Concept Statement' on 'activist anthropology' as formulated in Austin (to my knowledge the academic institution most explicitly proactive on these questions, as regards the study of Latin America), clearly connects these transformative and collaborative elements: 'Activist research begins with an alignment with an oppressed, disenfranchised, or marginalised collective subject, whose members are engaged in struggle for relief from oppression, for right and betterment'. This collective subject is to be involved in all steps of the research process, from the conception of the research questions and design, the data gathering, the analysis and the dissemination of results (Hale, 2001; Austin statement, 2003).

In activist proposals of anthropology, political responsibility and engagement are not seen as a 'corrupting' factor that would compromise the quality of the research, but on the contrary are given key importance in the research process. The political preferences and motivations of the researcher and the political situatedness of the researcher in the field are subjected to critical reflection. Charles Hale states in this regard that activist research involves 'making our politics explicit and up-front, reflecting honestly and systematically on how they have shaped our understanding of the problem at hand' (2001: 14).

We may understand activist research as an attempt to find satisfactory answers to the challenges and contradictions of the fieldwork experience. In their fieldwork, anthropologists are and have been directly confronted with poverty and suffering, with human rights abuses and violence, and have felt obliged to ask themselves what as human beings and *as anthropologists* they were doing about this.[14] Furthermore, anthropologists have increasingly been criticised for their lack of responsiveness and responsibility toward the communities researched and the failure to feed back their findings to their research subjects.[15] It has to be recognised that any anthropologist needs to gain a certain degree of acceptance and support from local actors, without which he would not be able to be present in their midst, to survive physically or to establish any form of meaningful interaction. To gain such legitimacy, the researcher needs to appeal to issues of interest to the local community and is likely to have to adapt his research to some degree to local demands and ideas. Thus, whether they wanted or not, anthropologists have long had to accept that local populations condition their work to a certain extent. The strength of collaborative anthropology is that it not only makes this aspect visible but theorises it and places it at the heart of the research endeavour. A particular strength that I see in activist and collaborative anthropology is that there is serious debate about the analytical consequences of this research practice.

[14] Hence, commitment and responsibility are not seen as just the *private* responsibility of the anthropologist as a *person*, but lead to a rethinking of the anthropological endeavour as such.

[15] See, for example, the work of Kay Warren, also discussed in Stephen (2002).

Rather than treating the influence of the needs and demands of research subjects as an unavoidable distortion or a necessary price to pay for ethnography that needs to be kept out of the analysis for fear of being considered lacking in academic rigour, collaborative anthropology instead claims that the involvement of local actors should be maximised to produce not only more socially relevant but also better, more adequate knowledge. Advocates of collaborative approaches expect the active dialogue with a collective subject to transform the research process, to make research more relevant to real-life problems and to produce a richer analysis. In their view, analysis is enhanced by the fact that as the research subjects have a greater stake in the production of knowledge, they will tend to be more committed to the research process. The same is true of the researcher who will be more aware of the possible consequences of his or her work. Furthermore, in collaborative work the need will arise to address differences and possible contradictions in the viewpoints between the collectivity and the researcher. It is precisely this exercise of engaging in a dialogue on these differences that brings the understanding of the social processes studied a step further (Hale, 2001; Earle and Simonelli, 2005). To my mind, the ways in which the so-called research subjects are engaged in a dialogue over the meaning and set up of the research is one of the key issues for contemporary anthropology. We need to come up with convincing and constructive answers to the question, posed by the populations we study, 'why should this interest us?' and recognise that, though this is by no means a simple exercise, the research has a lot to gain from it.

Authors like Nancy Scheper-Hughes place somewhat less emphasis on the collaborative aspect as such and relate militancy more to the needs to be addressed (e.g. stopping human rights abuses) or the values to be defended (e.g. human dignity). This latter position could, in theory, allow for ways of working that do not involve an alignment of the researcher with a specific and more or less formally organised collective subject. In practice, however, much of the research inspired by the activist or politically committed approach that I have encountered in Chiapas, seems to take the form of a collaboration with a specific social actor. I see this collaborative aspect as one of the central challenges to ethnography today and it will be this issue that I will discuss in relation to my own experiences.

7.3 Studying political conflict in Chiapas

Between 2003 and 2006 I developed a post-doctoral research project that looked into what I considered to be one of the key dimensions of the ongoing conflict in Chiapas, Mexico: the rivalry and coexistence between state governance and Zapatista autonomous structures at the local level (Box 1). The Zapatista autonomous structures of governance had been consolidated as a challenge to the Mexican state, when formal recognition of indigenous rights at the national level was unsuccessful. These structures consists of autonomous municipalities (*municipios autónomos*) and as of August 2003, the so-called *Juntas de Buen Gobierno* (JBG), Councils of Good Governance, bodies coordinating the autonomous municipalities. These administrative and political structures shape the everyday life of

Box 1. The Zapatista uprising

On the morning of January 1st 1994, the Mexicans woke up to a new reality. During the night, armed and masked rebels from the *Zapatista Army of National Liberation* (EZLN) had occupied a number of important towns in the southern province of Chiapas and declared war on the Mexican government. The (badly armed) army of indigenous peasants and their charismatic leader Subcomandante Marcos immediately made headlines around the world, forcing the Mexican government into a ceasefire. Peace talks stalled, giving way to a low-intensity conflict that is ongoing to this day.

Zapatista civilians, affecting their access to natural resources, their livelihood options, the way they relate to the military command structure of the EZLN, and the way they relate to other social actors locally and globally. Elsewhere I have provided a more detailed account of the ways in which these structures emerged and were consolidated and legitimised and I will not repeat this here (van der Haar, 2004, 2005a). A few things are important to bear in mind though for the present discussion:

- The Zapatista structures operate in parallel to the state system and challenge the state in its legitimacy and actual capacity to administer resources and people.
- The autonomous municipalities and other Zapatista structures are not formally recognised by the Mexican state and operate at the margins of the law. They are framed by the Zapatistas in terms of 'resistance'.
- The two kinds of structures overlap spatially and socially as groups of the (rural) population within and across communities align with either 'the government' or the EZLN. This gives rise in practice to a certain competition for exercising control over people and resources.
- There are no regions in Chiapas where the Zapatistas are hegemonic; they operate in contexts where a variety of rival political projects are present.

In my research I was interested in exploring local-level contestations of the state and changing notions and practices of governance, which I considered one of the most fascinating aspects of the Chiapas conflict. I was interested in contestation and change not only at the discursive level, but also at the level of practice, an issue which had been explored far less (Box 2). My field research zoomed in on the question as to how the two competing claims to governance of the municipal government and the Zapatistas interacted (the kinds of accommodations, negotiations and conflicts that were involved), what this meant for the way the Chiapas conflict unfolded in daily life, and the ways in which, in this context, local forms of governance were being reworked with reference to debates on indigenous rights, political participation and democracy. The project involved an eight-month period of fieldwork in the region of Altamirano, at the fringes of the Lacandón Rainforest. This was traditionally a marginalised region with a weak municipal government and also one of

Box 2. Competing claims to governance in practice

Picture two primary schools in one village of 500 inhabitants, one showing the slogans of the government, the other presenting the colourful graffiti of the Zapatistas. This is one of the most visible aspects of the rivalry between 'the government' and the EZLN in the rural villages of Eastern Chiapas. In the first, teachers paid by the state teach the standard Mexican curriculum; in the second voluntary 'educators' offer an alternative, revolutionary curriculum supported by national and international NGOs. The segregation continues after school hours, when young people of the village play basketball on either one or the other of the basketball fields near the schools. A more invisible rivalry relates to dispute settlement and local justice. Where the Zapatistas resolve their 'problems' at the level of the autonomous municipality or the Junta de Buen Gobierno, the non-Zapatistas turn to the judicial authorities in Altamirano. When a dispute involves members of both parties, negotiations follow between both sets of authorities.

the main centres of gravitation of Zapatismo. Several autonomous municipalities existed, (partially) overlapping with the official municipal territory as well as a regional centre (seat of the *Junta de Buen Gobierno*, also called Caracol) situated at only nine kilometres of the central town of Altamirano. Though Zapatista strength had greatly declined since its heyday in 1996/1997, it was still considerable at the time of field work. My main methodology was a combination of formal and informal interviews and participant observation and, especially, the following of specific 'critical events' occurring at the intersection of Zapatista and municipal governance. I expected to encounter such events which would allow me to analyse 'how authority is claimed, endorsed or contested, what institutional arrangements and normative frameworks are drawn upon in this process, how they are brought to bear upon one another and what the consequences of their interaction are.' (Nuijten *et al.*, 2004: 108).

Regarding how to go about my research and try to involve local actors, I had two, somewhat contradictory, ideas. On the one hand, I wanted to gain a closer involvement with Zapatista authorities at the level of the JBG in order to explore more closely how the widely known (and very appealing) Zapatista proposals on governance were given shape in practice and what their analyses and experiences were with regard to the uneasy coexistence with the official municipal government. On the other hand, however, I was also keen to study the situation of contested governance from the perspective of multiple actors, not only Zapatista authorities, but also official municipal authorities, people adhering to either 'the government' or 'the Zapatistas' or seeking neutrality. In this way I hoped to understand, from different angles, the dynamics of the conflict and the practices of governance at the intersection of the competing spheres. I was not entirely clear, from the outset, about which of these entry points to the field situation was my priority. However, I knew I was not prepared to accept working with the Zapatistas entirely on their terms and needed to keep doors open to other actors. Thus, my starting position was different from what collaborative anthropology advocates, where the alignment with one particular actor is taken as the basis.

My idea to try and analyse the dynamics of competing claims to governance from the perspectives of a variety of actors sprang from my concern with the apparent lack of such studies in recent scholarship on Chiapas. Despite the overwhelming scholarly attention to the Zapatistas and the Chiapas conflict, relatively few studies contextualised Zapatismo in relation to local histories and the interaction with other political actors. Neither an exclusive focus on the Zapatista ideology and strategy, nor the focus on other political actors while discrediting or discarding Zapatismo as irrelevant, seemed to me to afford the much needed insight into the ways in which the conflict was unfolding at specific sites, how the Zapatista project was being created and contested at different levels, and how it shaped the options, interactions and imaginations of both adherents and opponents. I identified with Xóchitl Leyva's plea to acknowledge local complexities and extend 'the analysis to the practices, discourses, and ideologies of all those involved' (Leyva, 2003: 182), given that in the case of Chiapas too often 'the Zapatistas' are juxtaposed with an undifferentiated 'rest' (cf. Van der Haar, 2005b). I did not, I should add, assume that the Zapatista structures were under assault from a more powerful local government. Rather, it was an open question to me, which I wanted to explore during the research, in what ways the municipal government dealt with the Zapatista presence, whether and how they placed pressures on them or tried to undermine them. Similarly, I was interested in the ways the Zapatistas might be exerting pressure on the municipal government. My assumption was that the workings of power needed exploration rather than viewing the Zapatistas as the *a priori* more vulnerable party. During the fieldwork and since then I have been wondering whether my decision to study multiple perspectives rather than identifying with one of the actors in the conflict (the Zapatistas) was a wise one. It certainly was not an easy option and produced continuous doubts and feelings of anxiety. I was often unsure as to whether my moving between different social and political actors – as I ended up doing- was not seen as a 'transgression' that would jeopardise my presence in the region (fortunately, this did not happen). Furthermore, as I will discuss in more detail below, my approach implied that the more intimate arenas of power on either side remained closed to me.

I did not have a very clear strategy as to how to approach the actors with whom I would need to negotiate formal permission for my research, the Zapatista regional authorities of the *Junta de Buen Gobierno* and the municipality of Altamirano. I hoped I would be able to establish a more or less formal connection with the Zapatista authorities given that I had good connections with the leaders of one of the autonomous municipalities on the basis of previous research work, but I did not know how much leverage they had at higher levels of the Zapatista hierarchy. Also, I was not at all sure how they would react to my idea of also interviewing people at the Altamirano town hall (*presidencia*). Similarly, I was not sure how the municipal government would react to my intention to talk to them *and* to the Zapatistas.

Next to formal interviews with authorities on both sides, I relied on finding enough room for informal conversations with people of different political affiliations affected by the competing claims to governance.

In the light of the proposals for collaborative anthropology discussed above, my approach may sound very naïve and improvised, but it was the way I had been doing my earlier research as well. I had not considered it very problematic until then since I had eventually always found enough opportunity to access the actors that I considered important. I was building, as before, on the contacts I had from the time I was a volunteer with one of the teams of the San Cristóbal Diocese and that I had maintained and deepened during later stays (also dedicated to research). I had grown into a habit of organising things alongside my research that would (I hoped) be of benefit to the local community, but my research itself remained mostly motivated by academic interests. Unlike many other researchers in Chiapas, who out of conviction or necessity work by aligning themselves in more or less permanent ways with peasant organisations or NGOs, I had been reluctant to develop such ties and negotiated my access directly at the community level.[16] I was well aware, however, that my previous ties to the Diocese continued to play a role in the way I was being viewed.[17]

7.4 Searching for dialogue

Who knows what they were thinking, the Zapatista representatives that received me at the regional administrative centre, the Caracol of Morelia, where I had come to present my research proposal to the *Junta de Buen Gobierno* (Box 3).[18] The treatment was polite and even kind. They apologised for having kept me waiting. I had not minded the waiting. It had given me the chance to see something of the dynamics of the Caracol and to greet some people I happened to know who were working there. It was a quiet day without much activity and I had plenty of time to think back to a previous visit to this place, a year or so earlier, when befriended Zapatistas had taken me along to a party there. I was impressed by the new buildings that had been constructed. When they were ready to receive me I thanked my interlocutors for the opportunity to present my research. After going over the formalities, such as taking down the data from my passport, they let me explain my intentions and asked

[16] I did have an affiliation, though, as a visiting researcher to an academic institution, the CIESAS Sureste in San Cristóbal de las Casas, which I myself considered very important but which did not, at the time, have any great significance beyond academic circles.

[17] I was well aware since my time with the Diocese that 'access' to the community needed to be granted through formal permission of the communal authorities and that without such permission physical presence in a community was not possible. Hence, in order to be able to talk to people in their home communities and share in their activities, it was necessary to negotiate access. This is a condition confronting researchers and other types of external actors all over indigenous Chiapas that has shaped the ways in which anthropologists have organized their research. (See also the introduction to Van der Haar, 2001.)

[18] To be precise, I was attended by two sets of representatives, firstly the Vigilance committee in charge of receiving the visitors and secondly by the JBG proper.

Box 3. Junta de Buen Gobierno

The installation of the Juntas de Buen Gobierno, in August 2003, brought a number of changes in the civil governance structure of the Zapatistas. One of the main objectives was to reorganise and regulate the interface with civil society. Uncoordinated support had produced inequality and friction. Furthermore, the JBG were entrusted with the coordination and vigilance of the autonomous municipalities. They were to prevent abuses of authority and make sure the principles of good governance were respected. Their functions also included mediating in conflicts between autonomous municipalities (also Van der Haar, 2004).

me questions for clarification. They listened but they did not react to questions from my side. My attempts to engage them in a discussion were unsuccessful. When I asked, after my introduction, and insecure as to what they made of it: 'What do you think? Do you think this could be an interesting issue?', some people seemed inclined to react but then held back. I then realised that I was not supposed to ask any questions and that we were not going to have a conversation. I had hoped for an opportunity to exchange ideas about the Zapatista challenge to the state and why I thought this merited research, but I realised this was not going to happen, at least not in that setting. I had also made a point of my disposition to take up any other issue that was more in line with their interests, offering my services so to speak, but this also elicited no reaction. The whole meeting produced a strange sensation in me. Most of the members of the JBG were very much like the people from the villages with whom I was accustomed to relating yet the dynamics of the interaction were so different from what I was used to.[19]

I was kindly given a date at which I could come back to inquire about the decision concerning my request. My proposal would have to be judged by the general assembly, I was told, but they would first have to decide whether to propose it at all. When I visited the Caracol again on the date I had been given, I learnt that somehow my request had not been followed up, but there was a record of my previous visit and I received an apology. I had a second interview, to be notified a few weeks later that they could not give me permission for the research because it was 'not in the people's interest'. I could hardly blame them and understood very well that they had other worries than my research. I said goodbye expressing my hope that in the future we might be able to exchange ideas.

The 'no' at the level of the Junta de Buen Gobierno came as no surprise. I had first explored the possibilities for research and the options for proposing my project, with the people I knew at one of the autonomous municipalities. They had made it clear to me that any

[19] The JBG was made up of representatives of the councils at the level of the autonomous municipalities who integrate the JBG in turns, on the basis of a complex rotation system.

research would have to be approved first by the JBG, even if I were to limit it to the level of this one particular autonomous municipality.[20] I was aware that, at the time of my fieldwork, the Zapatista organisation had just entered a process of what we might call rationalisation and a degree of centralisation. Autonomous municipalities were embedded more clearly than before in a hierarchical structure (linking them to the central command of the EZLN), and the responsibilities and competencies of the autonomous municipalities became more strictly circumscribed. A number of issues previously addressed at the level of these municipalities, or by individual communities, were now made the concern of the *Juntas de Buen Gobierno*, the regional centres coordinating the municipalities. With these changes, in practice the JBG started to act as a gatekeeper not only for NGOs or solidarity groups interested in supporting the Zapatista bases but also for researchers. Previously also, research projects (including those of students) needed to be negotiated and gaining permission could be complicated, but under the new conditions the room for manoeuvre at the local level was much reduced (at least, this was the situation at the time of fieldwork; it may have relaxed since). Upon my arrival in Altamirano, I had heard numerous examples of people (NGO workers and researchers) whose proposals for projects or research had either been rejected or not considered at all. I thus knew that the chances of success, at least in the timeframe I had, were reduced.

My interlocutors in the autonomous municipality explained to me that a policy concerning research projects was in the making in the JBG, in response to the many requests of students and researchers they were receiving. They also told me that there was a lot of hesitation about research projects, because, as they told me 'the people have started to analyse: the student gets the thesis but what do we get?' And, asking me why research about the Zapatistas would actually be necessary, they suggested 'Do not the *communicados* [the official Zapatista communiqués] explain it all?' In continuation, they also expressed some doubts about research because it can lead to 'criticisms' of the movement.

In this initial conversation, at the level of the autonomous municipality, I did not present very definite ideas for research and instead tried to explore whether there were interests they had that could match my own. I felt that the research question I had started out with was broad enough to accommodate anything related to Zapatista autonomy that they felt they wanted to address. But I found that the people I was talking to had a lot of needs and interests related to their situation as Zapatistas (some of which they shared with me) but few of these were 'research questions' or easily translatable into them. I realised once again how difficult it can be to translate academic proposals into words that make sense at the level of the people living the reality you want to study. I tried, of course. I talked about the fact that so little is known about the degree to which the state authorities respect the autonomous

[20] They in fact advised me against focussing on one autonomous municipality in particular because this was not "well accepted", since it was, at the time, the JBG that was supposed to allocate any type of project to one of the different autonomous municipalities.

structures and that many people worry whether this situation of 'two governments' (*dos gobiernos*) did not involve many conflicts. I was happy to find, at least, that the idea of '*dos gobiernos*' resonated well with the way they themselves conceptualised the situation and they were forthcoming with examples of conflict but also coordination between the Zapatista structures and the 'official' municipality of Altamirano. This is also an example of the rather informal and implicit ways in which I sought to validate the notions I was working with and refine my analysis.

In hindsight, I realise that I was focused on issues for *research* whereas their concerns were more in the nature of how to improve something or overcome an obstacle, for example, how to make headway with education for adults (I offered my services). I can see how a more change-oriented, collaborative approach would have meant engaging with the process of constructing autonomy in which my interlocutors, Zapatista municipal authorities, were involved, and with their searches and struggles. Any research question that would also be meaningful to them, would have to develop from there. Research would then be connected to efforts to improve something and its relevance would be clear. However, the starting point would then be to support an ongoing process rather than research per se, which is precisely what action-type research is advocating, of course, but which had not been my starting point.

Researching the Zapatistas is not impossible. I have met a few people, anthropologists and others doing their PhDs, who were able to do research with the approval of the Zapatista authorities and in good collaboration with them. Without exception, these were people who had a track record of work in support of the Zapatistas through an NGO or solidarity group. This suggests that a collaborative research project needs to be embedded in and follow from earlier engagements that give the researcher legitimacy, trustworthiness and insight. This earlier engagement is unlikely to be centred on research primarily, but rather on supporting the projects the organisation is trying to realise. Though I also had an earlier engagement in the region and had constructed a certain legitimacy, I had not built up credentials as a Zapatista activist and this proved crucial when dealing at the level of the JBG. My experiences also indicate, I think, that the extent to which research becomes collaborative depends in part on the researcher (personality, antecedents, time pressures, etc), but also on the particular dynamics of the organisation one intends to work with at the time one intends to do the research. I can imagine that the type of dis-encounter I had is as much part of the efforts of giving shape to a collaborative anthropology as are the rich encounters that make it so valuable.

Anticipating the official Zapatista 'no' and feeling a considerable time pressure to make some progress on my research, I had started to approach the municipal government of Altamirano. Here access was much easier, though also here some of the most crucial people were hard to access. But I was given access to documents and got the opportunity to conduct interviews. Though I had been open with the Zapatistas about my intentions to also know the views of the 'official municipality', as they call it, I suspected this move probably did little to improve my chances of getting permission from the Zapatistas. The Zapatista 'no' was a complicating factor in the research. It meant that I could not have access to the people or participate in discussions at this crucial level in the Zapatista governance structure. Nor could I formulate a joint project with one of the autonomous municipalities as I would have wanted. I could only informally communicate with Zapatista civilians about their experiences with the situation of 'two governments' (and I did this a lot). Therefore, I need to be very cautious when writing about the Zapatista perspective because much of it was told to me 'off the record'. Finally, the situation made it harder to access the deeper layers of understanding amongst Zapatistas.

7.5 Privileged perspectives

One of the questions about the activist and collaborative proposals that I am finding hard to resolve is: Does one, when doing anthropology the collaborative way, become too bound by the perspective of the organisation one is aligned with? And: how much of a problem is that? Studying 'from' or 'with' an organisation need not mean, or should not mean, a-critically adopting their perspective of reality. Charles Hale underlines that collaborative research does not mean that one is required to 'neglect alternative or contrasting perspectives' nor does it 'prevent stepping back to take in the big picture' (2001: 14). In practice, however, it seems to me rather inevitable that the researcher gets so absorbed in the outlook on things of the people he or she is surrounded by that it is increasingly difficult to think outside of this frame (unless perhaps a strong divergence arises, which in turn might result in ending the collaboration). These are the people the researcher feels loyal to, that legitimate his or her presence in the field, and it is their perspective the researcher increasingly empathises with as the research develops. The problem is not so much, in my view, that the researcher comes to adopt much of the value judgments of the organisation, but rather that it will lead him or her to dismiss other perspectives as either wrong or irrelevant. Ultimately, my concern is with researchers no longer being able to question dominant narratives about the organisation and its political rivals, to the expense of both dissident visions within the organisation and competing narratives about social change outside of the organisation. By opting to study multiple perspectives, this is what I tried to avoid.

There may be practical obstacles to accessing the other perspectives that one needs to get a sense of the bigger picture and that might help one to recognise some of the biases in the perspective of the organisation. Especially in a context of conflict, the opportunities to hold discussions with social actors from outside the organisation one has aligned with may be rather limited, precisely because of strong us/them constructions and enemy images turning any involvement or interest in 'the other' into a transgression and a security risk. In some ways I saw these dynamics of limiting the perspective reflected in Chiapas. Most of the other anthropologists I met were very sensitive and conscientious people whose work I very much admire. Yet, I noticed that those of them who were most embedded with 'the Zapatistas' had a tendency to limit their interaction with local non-Zapatistas. This was partly because these were outside their field of vision, though what also seemed to have played a role was care not to put at stake the confidence they enjoyed with the Zapatistas by transgressing local dividing lines. Similarly, I noted in some a tendency to think about 'non-Zapatistas' in rather black-and-white terms, without much interest in the, often very nuanced, critical analysis other social actors in Chiapas made of the Zapatistas.

Advocates of the collaborative approach are right to counter similar concerns about privileging the perspective of the organisation by arguing there is never one unquestioned and unquestionable organisational discourse, but that instead organisations harbour many different viewpoints. They are equally right to point out that many organisations are very capable of self-criticism. The advantage of the collaborative way would be precisely that the researcher can explore these different viewpoints and enter into these rich internal debates in ways that an 'outsider' never could. In the same vein, the researcher would be involved in the ways in which encounters or confrontations with other social actors, including political rivals, take place and share in the analysis that the actors he or she is involved with make of these. I can certainly imagine that this provides a richness to the analysis that would be difficult to reach in any other way than through a collaborative approach. Activist scholars argue that contradictions between the researcher and the collective subject they work with inevitably arise during the research process (some of them specific of activist research, others more general to the discipline). These contradictions can teach one a lot about the very processes that one is trying to understand (Hale 2001: 15). Confrontations generate new questions and perspectives and it is this which gives activist anthropology its analytical richness. I find this a convincing argument. However, I can also imagine that *whether* this richness is produced depends a lot on the organisational culture, the capacity for critical self-reflection, the extent to which dissident voices are seen as constructive, and on the leeway that the researcher is given to think and move beyond the established frameworks. As I suggested, in conditions of conflict, when the stakes are high and there is considerable polarisation, there may be less room for crossing the political boundaries and for self-criticism and plurality. This is of course a question that needs to be explored and on which I hope future collaborative endeavours will reflect.

As I mentioned, in my own research I made a more or less conscious choice to try and explore multiple perspectives on the situation of competing claims to governance. From what I know now about collaborative anthropology I am not sure that I would make the same choice, but I am also not completely sure that I would not. I certainly recognise the merits of the collaborative approach. I experienced that my contact with the different actors – and hence, on certain issues, my analysis- was more superficial than it might have been had I managed to come to a collaborative set-up. I expect that a collaborative approach would have given my interactions with people the stronger sense of urgency that comes with having a shared goal. Nonetheless, I had many valuable conversations in which people shared insights and concerns with me, giving me a good insight into how the competing claims to governance were experienced and analysed at the grassroots, by Zapatistas as well as others. What I missed most, however, was the access to the most crucial arenas and the more powerful people on both sides. Only indirectly, by piecing together many partial accounts, could I construct a picture of how the relationship between autonomous and municipal structures was negotiated at the higher levels of authority. I became quite convinced that accessing the more intimate spaces of power can only be done by a more 'embedded' approach along the lines of collaborative research. In the set-up I had developed, I remained too much of a free-floating individual to gain legitimacy at the level I would (also) have wanted.

Notwithstanding all this, I still think the efforts to capture multiple perspectives was valuable and that our scholarly understanding of Chiapas would benefit from more of such efforts. Many people I spoke with, both Zapatistas and non-Zapatistas, showed a very nuanced understanding of the complex dynamics of the conflict, more nuanced in fact than many scholarly analyses. Moving between different actors and different spaces was very enlightening in terms of finding out about the nature of the conflict in the region of Altamirano. First of all, the effort of moving between these spaces taught me a lot about where the dividing lines are between Zapatistas and others, how conflictive the situation was and where tensions occurred, but also where the dividing lines disappear and the relations become de-politicised. An interesting discovery was that much of what I feared could be seen as 'transgression', crossing symbolic dividing lines, was in fact much more 'normal' behaviour than I could have guessed. Furthermore, I learnt about the injustices some people felt they suffered at the hands of the Zapatistas, but also about the ways in which they identified with some of their proposals: their emphasis on human dignity and on the accountability of leaders, for example. I learnt a lot about the different positions, experiences and projects previously hidden under the broad label of 'non-Zapatistas'. Similarly, I got a quite differentiated idea of the ways in which Zapatistas experienced the current situation and how they thought about resistance and the future of the movement.

The question is, of course, whether I would not have been able to gain these insights equally or better through a more collaborative set-up. I am not sure. I do think however we need to reflect more explicitly on the consequences of the collaborative way of working for understanding diversity and contestation within and between collective subjects. I continue to believe that in the study of conflict it is crucial to acknowledge the existence of multiple, possibly divergent perspectives, not *a priori* discrediting any of these, and being attentive to the dynamics by which some narratives become dominant and others are silenced. I also believe it is crucial to suspend judgment about who is the more powerful or exploitative party and who are the victims to allow for more complex, harder to decipher workings of power.

7.6 Questions of commitment

In my understanding, collaborative and activist anthropology have the researcher's commitment to change as a starting point. From that follows the commitment to particular collective subjects working towards such change. Collaborative and activist approaches thus require the researcher to be clear about *what* and *who* he or she is working for. When I developed my proposal and the research on competing claims to governance in Chiapas, I did not deal with that question as much as I should have. Thinking back, I was motivated in the first place by a fascination for what was going on in Chiapas and what the Zapatistas were trying to achieve. I saw in the Zapatistas a movement that claimed to re-invent politics from below and to offer an alternative for the state in terms of services and security. I was (and am) intrigued and inspired by this endeavour. At this level, my commitment was with showing the capacity of indigenous people to innovate governance from below, to create security and progress, hoping in this way to increase the legitimacy of their challenges to the state and their proposals for reform.

Digging deeper, my motivation was to try and understand better what Zapatismo meant in the everyday lives of both the Zapatistas and those who are their neighbours. I was moved by a concern for the daily hardships and contradictions that civilian Zapatistas suffer. I had seen that the Zapatista movement was only partially able to fulfil the needs of its population in terms of education, justice, development and security. The strategy of resistance was costly for people in many ways and demanded time, commitment, and resources and I found myself wondering what this hardship was good for. I had problems with what I saw as the strong (perhaps increasingly so) centrally directed nature of the Zapatista movement. In contradiction to the strong egalitarian and participatory discourse, there were indications that the parameters of Zapatista internal social organisation and external contacts were set in a top-down manner. Also, an increasingly exclusive, dominant narrative about the uprising was being produced ruling out the diversity of voices that had existed in the beginning. I did not want to be complicit in legitimating this dominance with my work. My concern was with the Zapatistas at the base, who bear the cost of resistance, and my question was whether they were finding enough room within the movement to negotiate upwards, to have their concerns influence the course of the movement, to organise things differently and to construct alternative meanings of Zapatismo. My commitment, ultimately, was with the Zapatista bases and their neighbours, and the hardships of marginality, political exclusion and conflict, and with uncovering their narratives. If I hoped to support a process of change with my research, it was related to giving legitimacy to the experiences and narratives of these people at the basis, to creating recognition for Zapatismo as a multivocal project that is constructed at multiple levels and, ultimately, to creating more room for negotiation from below.

Finally, my commitment was also with overcoming dichotomous images of a Chiapas divided into Zapatistas and non-Zapatistas. I believed it was important to recognise the multiple interactions and inter-permeability between different groups in local society. Many erstwhile Zapatistas had left the movement but still felt connected to it. Though the EZLN had made efforts to clear up the grey areas between being yes or no a Zapatista, local people constructed different, more fluid, layers of loyalty. Furthermore, the situation of competing claims to governance was affecting not only the Zapatistas but also their neighbours. I believed it was important to understand how different groups were affected and how they moved in that landscape of rivalry.

To serve these purposes, I considered it crucial first and foremost to conduct a conscientious and perceptive ethnography. Or, as Nancy Scheper-Hughes put it: 'Seeing, listening, touching, recording can be, if done with care and sensitivity, acts of solidarity. Above all, they are the work of recognition. Not to look, not to touch, not to record can be the hostile act, an act of indifference and of turning away' (1995: 418). However, this commitment with minority voices, with the groups that are silenced or marginalised in the political rivalries of Chiapas and the effort to make their experiences more visible, in some sense lead me away from the collaborative approach. This commitment was not easy to 'match' with one particular collective subject. In hindsight, it is perhaps not surprising that I did not manage to convince the Zapatista authorities of the use of my research, because I was not fully identifying with what I saw as the dominant Zapatista project. The question is really whether, or under what conditions, a collaborative set-up can give shape to a commitment with minority voices and understand the internal power dynamic. Such an enterprise might easily lead one to 'undermine' the image that the organisation, in this case the EZLN, is trying to construct of itself.[21] In a period when a movement is coping with a number of difficulties, this can be very threatening. I am inclined to believe that collaborative research can only really work in those situations where there is openness and willingness to acknowledge the dissident voices within a movement or collective and to critically examine the internal workings of power (of course, I cannot be sure such a willingness was indeed absent in the case of the Zapatistas because we never even got close to exploring that question).

However, to be fair to the collaborative proposal, I must also admit that the concerns I formulated were *my* concerns. The concerns with the costs of resistance and with the lack of room for bottom-up construction of Zapatismo were based on observations and conversations I had with numerous individuals, in and around the movement. But I never really engaged with the question to what extent these were also the *central* concerns of the civilian Zapatistas I was so worried about. Was the construction of a centralised Zapatismo seeking to discipline its bases, as I saw it, really a priority problem that *they* felt needed to be solved? They were critical about these issues in some regards, and they showed anxiety about the high costs the movement implied for them, but I never discussed explicitly with them whether in their eyes this was an urgent need that should be addressed. Possibly, the internal tensions and the deprivations that were the consequence of the policy of resistance were felt to be of less importance than the survival of the movement as such. Perhaps, their foremost concern was also the need to overcome the practical obstacles to keep the autonomous structures going. I also concede in favour of the collaborative approaches that, had I had the opportunity, I would have learned a lot from researching from within, at different levels of

[21] See Hilhorst (2003), especially the epilogue, for an experience in which this proved impossible.

the Zapatista movement, how the construction of the agenda took place, what the points of divergence were and how these were negotiated, what actors and what narratives were being sidelined and why, and who experienced this as a problem.

Recognising the value of collaborative and activist approaches, I do however make a plea for also developing ways of responsive and responsible anthropology that are not strictly tied to a collective subject. In addition to the reasons outlined above, my concern is that a collaborative approach might turn out to be exclusionary because of the requirements it places on the nature of potential collective subjects to act as the researcher's counterpart. These approaches seem to need or presuppose collectivities that are relatively identifiable and structured. Constructing co-ownership of the research process is much more feasible, and perhaps hardly imaginable without, a collectivity that has an identifiable and accessible structure of representation, decision-making and membership. Without that, what would dialogue, participation, and accountability look like in practice? One of the consequences of this could be that the more visible collectivities, those that are well-organised and present a well-articulated discourse that we as researchers can relate to, are privileged over those sectors of society that are less articulate and less visible, both the marginal actors within these organisations and the less-articulate actors outside of them. I see this happening to some degree in Chiapas where the Zapatistas have attracted most research attention, followed by other progressive groups that defend similar values of human dignity and that are equally embedded in global solidarity networks. Of course, these organisations merit this attention. The aggregated effect, however, is that some voices in society, of the less organised poor, or of those that stand for other less 'appealing' political options, fail to become visible in the scholarly debate. These are perhaps voices that have not found an attractive language of human dignity, and that we do not recognise as embodying a possible alternative of social change. This may imply that we do not put our images of who is right and who is wrong, of who are the oppressors and who the victims, to the test and do not really engage with the fractured reality of conflict and exclusion. Overlooking the less attractive voices and more elusive subjects may also mean that some of the most dire human suffering remains invisible to us.

7.7 Conclusions

In this text I have confronted the proposals of activist and collaborative anthropology with my own research practice and have reflected on the promises and challenges of these approaches. Revisiting my own research through 'collaborative' eyes was both instructive and, at moments, painful as I realised the ways in which I had evaded some of the crucial questions about commitment and the relevance of my work. The exercise convinced me of the importance of collaborative work yet also helped me to identify more clearly the points on which I remain hesitant about it. The added value of collaborative work is evident, especially when the conditions allow for an open and constructive mutual engagement. However, these conditions are not always present. The researcher might not have the right skills, preparation or enough time and flexibility to seriously engage in the demanding and unpredictable dynamics of collaborative work. The organisation may not be, at that moment of its development or because of the context, willing or able to deal with critical debate. In practice, collaborative research in its fullest sense, might only be possible in a limited number of situations. As one of my colleagues put it, having done his PhD research in collaboration with peace building NGOs, one cannot be sure about the degree to which the collaborative intentions that exist at the beginning, come to fruition during the research journey (Van Leeuwen, 2008). To be sure, an experience in collaborative anthropology should not only be considered successful when the collaboration is accomplished in all its facets. Not only the encounters, but also the dis-encounters during the research journey, or even the abandonment of the joint research, can be very instructive.

My main hesitations with collaborative work are two-fold. First, I am concerned whether collaborative research is a good way to explore power dynamics within and between (rival) social or political movements and whether it might not, unwillingly, contribute to legitimising dominant discourses within organisations and thus contribute to exclusion and internal disciplining. Second, I am concerned with the question as to whether collaborative research, with its focus on collective subjects, does not blind us to the unarticulated, the politically incorrect or the hardly recognisable actors. These problems are related especially to the association of the researcher with a particular collectivity which is at the heart of the collaborative proposal. I would like to make a case for a diversity of ways in which to give shape to a responsible and responsive research practice, beyond such an exclusive association.[22] Diversity in the research questions that we pose, diversity in the ways in which we define our commitments, and diversity in the ways in which we engage with social actors in conducting research seem to me healthy and necessary in the development of anthropology as well as development studies. This diversity in approaches is important particularly when studying conflict. In situations of conflict, the association with one or another collective actor positions the researcher variously in the conflict dynamics with consequences for the kind of analysis that he or she makes. Studies where the commitment and collaboration with a certain group or organisation is a starting point and studies aiming to understand multiple voices and their interaction, both have important stories to tell about the conflict experience and process, and both should have a place in the anthropology of conflict. By raising these issues here, I am not suggesting that collaborative scholars are unaware or insensitive to them. Indeed, I am very much looking forward to new studies being published by collaborative and activist scholars which will no doubt address these concerns and many others.

This text was also written with the intention of helping the readers of this volume to design their own research. I end therefore by summing up some of the lessons I draw from my Chiapas experience. In my view it is very important for any anthropologist (or any academic

[22] See also Laura Nader in her reaction to the Scheper-Hughes 1995 article.

for that matter) to engage with the questions raised by collaborative and activist approaches, regarding: What moves you? What do you hope to contribute? Where do you see your responsibility? Who or what do you think you should be responsive to? The next question is then what options you have to shape this and in what kind of relationship with what kind of social actor. When deciding whether to embark on a collaborative set-up, the researcher should not only consider what suits her personality, background, motivation and financial and time constraints, but also the research setting and the nature of the organisation she would like to work with. Are there previous contacts with organisations that the researcher can build on? An important lesson from Chiapas is that a certain solidarity with what the organisation stands for and previous engagement involving other than only research activities, might be an important pre-condition to developing collaborative research. Even then, an assessment needs to be made as to whether there is a mutual interest in collaboration and whether enough mutual trust can be constructed to deal with sensitive issues and possible divergences. This takes time, skill and flexibility. The researcher therefore needs to ask himself whether he is prepared and able to enter into such an open-ended process over which he has only partial control. Last but not least, it is important that the institution within which the researcher works (the university or research institute) is supportive of collaborative work and willing to accommodate changes along the way and possible delays.

Acknowledgements

The research on which this text is based was made possible by a post-doc grant from the WOTRO foundation in the Netherlands.

References

Austin statement, 2003. Concept Statement Activist Anthropology at the University of Texas, Program in Activist anthropology. University of Texas at Austin version 23 October 2003, http://www.utexas.edu/cola/depts/anthropology/programs/activist/. Accessed May 2005.

Earle, D. and J. Simonelli, 2005. Uprising of hope: Sharing the Zapatista journey to alternative development. Altamira Press, Walnut Creek.

Hale, C.R., 2001. What is activist research? SSRC 2 (1-2): 13-15.

Hilhorst, D., 2003. The real world of NGOs: discourses, diversity and development. ZED books, London, New York.

Leyva, X., 2003, Regional, communal and organizational transformations in Las Cañadas. In: J. Rus, R.A. Hernández Castillo and S.L. Mattiace (eds), Mayan lives, Mayan utopias: the indigenous peoples of Chiapas and the Zapatista rebellion. Rowman & Littlefield Publishers, Lanham, pp. 161-184.

Nuijten, M., G. Anders, J. Van Gastel, G. Van der Haar, C. Van Nijnnatten and J. Warner, 2004. Governance in action: some theoretical and practical reflections on a key concept. In: D. Kalb, W. Pansters and H. Siebers (eds.), Pathways of development: Themes and concepts in current research.: Kluwer Academic Publishers, Dordrecht, Boston, London, pp. 103-130.

Scheper-Hughes, N., 1995. The primacy of the ethical: propositions for a militant anthropology. Current Anthropology 36: 409-440.

Stephen, L., 2002. Zapata lives! Histories and cultural politics in Southern Mexico. University of California Press, Berkeley.

Van der Haar, G., 2001. Gaining ground: land reform and the constitution of community in the Tojolabal Highlands of Chiapas, Mexico, Latin American Research Series. Rozenberg Publishers, Amsterdam.

Van der Haar, G., 2004. The Zapatista uprising and the struggle for indigenous autonomy. European Review of Latin American and Caribbean Studies 76: 99-108.

Van der Haar, G., 2005a. El movimiento zapatista de Chiapas: dimensiones de su lucha. Labour Again Publications, IISG, Dossier on rural and indigenous mobilizations in Latin America. Available at: www.iisg.nl/labouragain/documents/vanderhaar.pdf.

Van der Haar, G., 2005b. Chiapas and the Zapatistas: Filling in the picture (Review Essay). European Review of Latin American and Caribbean Studies 78: 101-106.

Van der Haar, G., 2005c. Engaged or activist research: review of a trend in Latin American Anthropology. Unpublished paper.

Van Leeuwen, M., 2008. Partners in Peace: Discourses and practices of civil-society peacebuilding. PhD thesis Wageningen University, Wageningen.

8. Deploying critical capacity

The case of sociology, experts and laypersons in the controversy over GM maize in Mexico

Gerard Verschoor

8.1 Introduction

I spent my childhood, adolescence and early adult life in Mexico. There, from the position of an upper middle-class white male, I often wondered about the ways of life of many of the people I came across on a daily basis (as an anecdote: Larissa Lomnitz's (1977) excellent *Networks and Marginality* – a book on poverty in Mexico City referred to a slum just down the road from where I lived). Injustice and poverty certainly left its imprint, and in retrospect it was no surprise that I chose to study (and later work in) rural development sociology. Although the choice to have done so may have come from the heart, rationally I guess the choice was made by the conviction that it was *the* thing to do to make the world a better place. At the time at least I saw development sociology as providing the adequate tools for the job to understand the underlying causes of injustice. As an extra, I benefited greatly from the opportunity to partake in Norman Long's (then professor at Rural Development Sociology) painstaking effort to bring flesh-and-blood actors back on the agenda of development research and practice (a difficult enterprise at a time when 'macro' approaches had the time of their life). However, it soon dawned on me that – despite the excellent studies on development interventions carried out at the time, see for example Long and van der Ploeg (1989) – the focus lay on the *interpretation* of processes of change. In other words, social actors did participate in Wageningen's 'actor-oriented approach', but only insofar as they could choose to learn the results of the studies or ignore them.[23] But social actors could certainly not add to the studies, dispute them, and even less contribute to their construction.

So it became evident to me that the role of laypersons in sociology was a subsidiary one. Whether it was about natural resource management, state or market intervention, or the impact of capital and technology on rural dwellers (the usual objects of development sociology), the 'actors themselves' remained silent. Searching for the 'deeper causes' of this state of affairs brought me not into contact with action research (a pariah in social sciences at the time) but with writings on the history, philosophy and sociology of science (e.g. Kuhn, 1970; Serres, 1997; Latour, 1987). At that time, there was one quote (in a book on the construction of science) which, in retrospect, set in motion a whole range of things: 'facts are like cows: look at them long enough in the eye, and they run away.' (Knorr-Cetina, 1981:

[23] Most often not even this choice was available since the studies in which they featured were only rarely published in the language they spoke.

3). From this quote it was only a relatively short step to the recognition that the exclusion of laypersons from science was more of a political machination than an inflexible 'fact of nature'. Academic requirements, teaching commitments and – perhaps most importantly – a lack of debate on the issue discouraged me from enquiring into the matter. Yet the work I was reading (mostly social studies of science stuff) helped me identify one of the themes where the issue of citizen participation was writ large: the interface between science and society. And so it was that by 2002, in preparation for a sabbatical leave, I chose to delve deeper into issues to do with biopower and, more particularly, GMOs (genetically modified organisms). This was a field where passions ran high and where science did not, or could not anymore provide all the answers. More interestingly, whenever scientists offered their points of view on issues related to GMOs (either in favour or against) these views were immediately contested – often by scientists from 'the other camp', but more usually by laypersons or by scientists-turned-activists.

Just such a controversy was unfolding in Mexico at the time. The discovery by Greenpeace, in March 1999, that genetically modified, unprocessed (and hence cultivable) maize seed was entering Mexico en masse attracted the attention of many but, as usual, was ignored in the best depoliticising tradition of the Zedillo Administration. The situation however imposed itself on the national agenda at the end of 2001 after two Berkeley-based researchers ignited a scientific scandal[24] of unprecedented magnitude by confirming suggestions that transgenic DNA had contaminated local maize varieties in remote regions of Puebla and Oaxaca (Quist and Chapela, 2001). The discovery and subsequent, 'scientific'[25] confirmation of genetic contamination was first taken up by a relatively small number of NGOs, concerned scientists and specialised journals, but from there it slowly moved into the mass media. Soon thereafter farmer unions, indigenous communities and a wide range of NGOs protested against genetically engineered maize, which itself became a high-profile topic within broader social movements such as *el campo no aguanta más*[26] as well as movements specifically created for the purpose of strengthening local maize (e.g. *en defensa del maíz*[27] *or the Foro Campesino por las Semillas y la Vida*[28]). And so it is that, by the end of 2002, a controversy had been

[24] The article by Quist and Chapela (2001) was first criticized by a Berkeley group of scientists opposing the findings. The main accusation was that Chapela – who backed the anti-GM campaign in Mexico – was a clear exponent of an activist-turned-scientist who had found what he desperately wanted to find. In an unprecedented move *Nature* - after listening to the arguments of this group – retracted the article. What followed was a heated exchange over conflict of interest and scientific discourse as the group critical of Quist and Chapela's study was found to have had their research funded by biotech giants such as Monsanto. See Scott (2003) for an account of this controversy, and of the breakdown of scientific discourse following this episode.

[25] I deliberately put brackets here because the extent to which scientific proof is *really* a proof (i.e. whether the methodology followed was 'good' or 'flawed') is itself a matter of hot debate.

[26] "The countryside can endure no more". See Bartra (2004) and the volume edited by Schwentesius *et al.* (2003) for an in-depth analysis of this movement.

[27] "In defense of maize".

[28] "Peasant forum for seed and life".

born. From the perspective of my personal background the controversy was a god-send, and so I chose to spend my sabbatical in Mexico, doing fieldwork on the topic right where the disputes were being deployed.

In Mexico I was hosted by a renowned agricultural research institute just outside Mexico City. I had chosen this institute because I expected it to be in the midst of the controversy. To my dismay this was not at all the case. Other than a relatively safe analysis from the sidelines, virtually no scientist from this institution intervened (either positively or negatively) in the controversy on GM maize.[29] After this initial disappointment I looked further into the Mexican scientific landscape for traces of scientific involvement. It did exist, but only on the natural sciences' side of the campuses (and they invariably backed genetic modification anyway). On the social science side, however, the attitude was always more or less one of denunciation in the manner of what Paul Ricoeur calls the 'hermeneutics of suspicion'. These studies lacked the opinions of those flesh-and-blood actors (and their spokespersons) directly affected by genetic modification of maize. The justification for this stance can best be described by the idea that those involved in the controversy were too busy to see what the issue was *really* all about and so it was the social scientists' task to uncover the 'truth'. After all, these social scientists had been trained to be reflexive and identify the forces ('capitalism', 'empire', 'fields of power' and so on) ordinary citizens could not see by themselves. A bit bored by these self-complacent stories I then set out to meet the 'actors themselves' who, I was warned, were 'deluded'. Contacting activists, going to meetings, interviewing key proponents (and opponents), participating at sit-ins I soon became a familiar face in the world of (anti)GM maize. Against all the advice, I aimed to do things differently: to learn from the actors in the controversy how I should talk about them, instead of telling them how they should think about themselves.

Upon my return to the Netherlands I put the material to rest for a while: a sabbatical is a nice thing to have, but 'normal work' had accumulated during my absence in the field so I could not publish anything on the subject until nearly two years later. In the meantime, this idea of how to talk about those actors (whom I learned had the same critical capacity and reflexivity as any social scientist) nagged at my brains. A further delving into the literature took me to (modern-day) pragmatism, especially the work of Boltanski and Thévenot (1991) and their idea that ordinary citizens are in absolutely no need of a sociological or otherwise explanation of what they are doing, or why.[30] It then dawned on me that – just like the actors whom I sympathised with, and learned from – I in fact was playing the same game: deploying arguments (maybe through other means, but nevertheless) through re-presentation, that is, by presenting things in a new light. My stories were thus 'one among

[29] To be fair I must add that the situation has changed since. Today a number of scientists from this institution (most of them from the natural sciences) are actively participating in the controversy – often by providing evidence against the negative externalities entailed in GM maize.

[30] For a fuller explanation of their position see the next section of this Chapter.

many'. They were nothing more, but also nothing less, than the stories or re-presentations of those involved in the controversy on GM maize. Armed with this insight I then decided to take a 'risk' and say something new (and not just a re-hash of what everybody already knows) about the controversy and, more especially, about the risk that activists' arguments against GM maize might in the end not be taken on board because of their propensity to be assimilated by those proposing GM maize.

The result is the story that follows, which is built up in seven sections. I first introduce the concept of 'regimes of justification' used by actors engaged in moral and political struggle. A second section displays the way in which arguments related to market and industrial efficiencies are strategically built and circulated by proponents of genetically engineered maize. This is followed by a paragraph on the broad coalition of farmer organisations, indigenous groups and NGOs opposing GM maize, who evaluate and justify their actions by deploying arguments related to equality, solidarity, culture, tradition, and the environment. In a fourth section I elaborate on the commonalities between the contenders – that is, the reliance on *certainty* concerning the relationship between means and ends – and argue that, as long as the anti-GM movement sticks to well-established, received frames of justification their struggle runs the risk of being futile. This part is followed by a focus on some of the uncertainties inherent in crucial economic and ecological processes pointed out by opponents to GM maize. Finally, and by way of conclusion, I indicate some of the broader effects this circulation of uncertainties may have in terms of effectively re-politicising the controversy about the protocol to follow concerning the place of GM maize in Mexico.

8.2 Disentangling the controversy: 'regimes of justification' and the common good

We live in a complex world. A world full of controversies. Controversies which seem to proliferate by the day, and affect us all. Many actors become – wittingly or unwittingly - entangled in these controversies. Importantly, traditional notions of the way in which controversies should be handled tell us that there is a specific category of actors (experts) who, because they are more knowledgeable, will find rational solutions. This traditional view also tells us that there are other actors (laypersons) who express their concern in any given controversy, but who nevertheless are thought to be less knowledgeable and hence in need of education by the experts. Finally, and as far as the role of sociologists is concerned, the traditional view tells us that it is their task to provide a neutral or objective commentary on their controversy of choice, and to do so in terms of an analytical metalanguage.

I have a problem with this rendering. Commonsensical and perhaps cherished notions are beginning to show cracks. This is evident in many of today's controversies concerning, for example, nuclear power, climate change, avian flu, and so on. In these controversies, the issues involved vastly overflow the knowledge of the experts who are supposed to contain these controversies. A hole in the ozone layer, for example, is not only a phenomenon that can be addressed by a meteorologist, but rather an event that simultaneously concerns chemists, politicians, oncologists, economists, lawyers, or people like you and me. Clearly, controversies of this type very much annoy the experts: they have been trained to 'purge' the phenomena of their choice from all possible 'impurities' (cf. Douglas, 1966) so as to make them amenable for study and, possibly, solution. Framing the controversies in ways that conform to their specialisms: that has been the task of the experts (and, it would be fair to say, a task in which they have succeeded very well!).

Many of today's controversies however resist framing or reduction in the traditional way. As the example of the hole in the ozone layer suggests, conventional ways of framing issues are now showing overflowings – economists call them 'externalities' – that are difficult to contain. Crucially, these new controversies are witnessing the emergence of a contestation of expert knowledge by laypersons. I suggest that this is so because these modern controversies touch on fundamental issues related to morality and ethics. As Latour (1999: 243) argues, 'if there is one thing that does *not* require an expert and cannot be taken *out* of the hands of [the lay people], it is deciding what is right and wrong, what is good and bad.' This is exactly what laypersons are claiming in many of the controversies that Callon and Rip (1986) have coined 'hybrid forums'. This concept is used to characterise highly confused situations (e.g. mad cow disease, the hole in the ozone layer, and so on) in which 'facts and values have become entangled to such an extent that it is no longer possible to distinguish between two successive stages: first, the production and dissemination of information and knowledge, and second, the decision-making process itself' (Callon, 1998: 260). In these forums both experts and non-experts actively participate in attempts at interpretation and evaluation, without however being able to reach closure. Experts, for instance, when faced with the proliferation of overflowings generally try their best to contain the arguments of laypersons. They thereby cling to former framings, or even try to strengthen and reinforce them. Laypersons, on the contrary, fathom the possibility of other framings (thereby often mobilising non-moral and non-ethical arguments such as scientific evidence that may jeopardise what the experts are saying!).

What role could the social sciences assume in this period of overflowings, unsettled controversies, hybrid forums? Should one be content with providing a 'neutral' commentary from 'above' or from 'outside' these forums? Or do we have to take a stand? It is my conviction that we have to do the latter, quite simply because we are all in the same boat. There are no longer any external 'observers' large-scale phenomena such as BSE, SARS, *r*BST, HIV-AIDS, GMOs. We are all participants in the same hybrid forums shared by experts and laypersons, and hence we too should have our say. So the predicament is this: either we sociologists are in the camp of the eliminationists and reductionists who try to purge Modernity from its impurities, or we join forces with all those who are trying to conceive of new protocols to address controversies in all their complexity. It is my conviction that it is more productive to do the latter. This poses a problem though, because sociology was not designed for this task. At least since Durkheim, sociology has been tainted by the incredible pretension of wanting to act as legislator (Bauman, 1992). That is, from its inception sociology has taken as one of its main tasks that of providing an explanation of why actors are involuntarily manipulated by forces exterior to themselves and known only to the social scientist's powerful gaze and methods (Latour, 1999). Thus a metalanguage was created (think of concepts such as 'structure', 'language', 'society', 'culture', 'power', 'capitalism' or 'fields') in which the language of ordinary, lay actors was thought to be embedded.

But what if it is not the social scientist's job to teach others how they must think of themselves? What if sociologists lacked the knowledge not only of what actors do, but also of why and how they do it? What if sociologists suddenly found themselves on the other side of the fence, having to learn from lay actors whether or not they were asking the right questions? What if it is not the sociologist's task to legislate the world, but instead capture its messiness, its creativity, and become involved in its (re)assembly? Letting go of the dubious insolence that sociology may see or 'uncover' invisible forces known to nobody but themselves would level a playing field that has been dominated by social science for too long. To be sure, much would be lost (including, of course, issues to do with legitimacy and scientific authority). Much more, however, could potentially be gained *by turning actors' practices into a resource* (and not a subject of analysis). In this Chapter it is my intention to develop only a small but important part of these untapped resources: that of the 'critical capacity' of lay actors (Boltanski and Thévenot, 1991, 1999, 2005). In a nutshell, these authors develop a framework designed to illuminate the most legitimate types of arguments – which they call *justifications* - that actors use in public disputes to assess what benefits the common good. Justifications can involve positive arguments, claims, or position statements, but might also be critical denunciations of opposing views. A justification in this view is 'an attempt to move beyond stating a particular or personal viewpoint toward proving that a statement is generalisable and relevant for a common good, showing why or how this claim is legitimate' (Thévenot *et al.,* 2000: 236). Actors engaged in argumentation are thus seen as mobilising so-called 'grammars of worth' or 'regimes of justification' – each with its own way of separating good from bad, right from wrong, just from unjust. For the purposes of this Chapter, the most important grammars or regimes include: 'market' performance; 'industrial' efficiency based on technical competence and long-term planning; 'civic' equality and solidarity; 'domestic' and traditional trustworthiness entrenched in personal and local ties; and 'environmental' friendliness. What is new in this model for the analysis of controversies is that the 'grammars' or 'regimes' of justification are quite independent from one another,[31] and each – and this is the great strength of the model – has the capacity to denounce the others because they lack morality or virtue (Latour, 1998: 224).[32] In the next two sections I illustrate these 'regimes of justification' by way of a summary account of arguments put into circulation by, respectively, proponents and opponents of GM maize.

---

[31] This is the result of a long history of political philosophy; see Boltanski and Thévenot, 2005).

[32] In this sense, then, justifications can be seen as 'strategic' because they entail the intentional construction or mobilization of arguments to be presented to the 'right' audience in the 'right' situation.

8.3 Justifying the common good: the arguments of proponents of GM maize

As we have seen in the previous section, actors deploy the 'strategic' practice of building and using arguments (justifications, denunciations) to forward their version of what is 'good' or 'bad' in relation to a common good (in this case maize). This entails an instrumental manipulation of arguments fitted to the controversy and aimed at advancing a strategic plan in the most legitimate terms that are culturally available (Boltanski and Thévenot, 2005). This section highlights some of the criteria that are utilised in this quest by pointing to a series of justifications (and denunciations) that are used by proponents and opponents of GM maize.[33]

Justifications based on the industrial world ('industrial worth')

Probably the greatest number of arguments in favour of GM maize are based on justifications based on this order of worth. In this category arguments, persons, activities and objects are positively evaluated when they contribute to a very specific end: that of efficiency and productivity.[34] Following Boltanski and Thévenot (1999: 373),

> 'In an industrial world the great persons are the *experts*. The words used to describe their personal qualities can also be used to qualify things. They are said to be worthy when they are efficient, productive, operational. They implement tools, methods, criteria, plans, figures, graphs, etc. Their relationships can be said to be harmonious when organised, measurable, functional, standardised' (my italics).

In Boltanski and Thévenot's framework, experts (professionals, specialists, and all manner of 'persons in charge') often, if not always, obtain their strength from scientific evidence and certainty. Below, I provide three examples of ways in which this is made apparent. The first one can be seen at play when proponents of GM maize argue that food security can be achieved by engaging in a game of 'comparative advantages' that will benefit all: proponents maintain that under these circumstances rational Mexican farmers will surely seek to grow those products that fetch a higher price in open, perfect markets while at the same time guaranteeing that they have access to 'cheap' maize.[35] This form of reasoning includes, by

[33] A word of caution is required here. The orders of justification presented below might give the impression that actors entailed in passive argumentation put on their blinkers and limit their justification to a specific order of worth only. This is not at all true, and evidence shows that actors are quite ready to mix several arguments, switch between them, reach compromises, or strategically focus on one type of justification to attract a specific audience. Versatility is thus key in strategic argumentation.

[34] Justifications in this domain are almost exclusively made by defenders of GM maize.

[35] Javier Usabiaga (Minister of Agriculture from July 2000 to September 2005) took this stance. His adversaries often criticized him for this, because it showed his "lack of responsibility" and "deficient social sensibility" towards farmers in danger of being squeezed out of the market altogether by cheap, subsidized agricultural imports from the U.S.

extension, trade in GM maize. The rhetoric used is clearly based on the universal applicability of a rational logic (comparative advantages) that can be accurately implemented by means of all sorts of tools, methods, criteria, standards, regulations and the like. Among the most visible we have the enforcement of intellectual property rights, and the harmonisation of biosafety and certification procedures and standards with those of the international treaties (especially those supported by the U.S.). In short: under the regime of industrial justification all organisational devices that are directed towards future, long-term planning and investment are seen to be worthy, while any move opposing this is seen and understood to be wrong, amoral or harmful.

The second example concerns the impact of GMO's on the environment. One recurrent issue here is the concern that genes from second-generation GMO's – such as pharmaceutical transgenic maize – might find their way to local landraces in, for example, Oaxaca. Here proponents of GM maize favour specific, scientifically 'sound' methods to control the risks involved. Read Monsanto's Robert Horsch on the issue of pharmaceutical crops during an international forum on gene flow held in Mexico City in September 2003:[36]

> 'What I can tell you for the short term is: the pharmaceutical production in maize is handled in an extreme different way than commodity crops. In fact, it is not handled in any way like a commodity crop. And it would not be possible to accidentally stumble upon it. And yet there are quite a few safeguards in place to prevent deliberate mischief to try and misappropriate it. It is grown in great isolation, with only specific contracted growers, under a highly regimented set of standard operating procedures and quality control methods which are required for producing any pharmaceutical, but which go orders of magnitude beyond conventional seed quality control systems, for example, to prevent unintended movement... And I'll stick my neck out here a little bit, and say that if somebody thought it ended up in Oaxaca, that we would take a very active interest in trying to track that down and put it to rest one way or the other. But by far the frontline of defence is systems that will prevent that from ever happening.'

What we can hear in Horsch's position is the unbridled optimism of a scientist who is certain about two things at least: 'biopharm' is a good thing in principle and – provided the right safeguards are in place – there is no risk that these transgenes might 'escape' the scrupulous and vigilant eye of 'the system'. In short: there is confidence that the means (a safety system) will be instrumental in reaching a stated end (efficient production of pharmaceuticals).

[36] At the time, Horsch was Monsanto's vice-president for international development partnerships. In November 2006 he joined the Bill & Melinda Gates Foundation as senior programme officer, focusing on improving crop yields in sub-Saharan Africa.

Justifications based on the market ('market worth')

Justifications in this domain are almost exclusively made by defenders of GM maize, and refer to a market or commercial logic. From this point of view, important persons are buyers and sellers who compete with one another towards an exclusive end: the acquisition of scarce goods. Within this frame of justification, the main qualities of these persons include their being 'opportunistic in spotting and seizing the opportunities of the market, to be unhampered by any personal link and to be emotionally under control. They connect with one another through competitive relations' (Boltanski and Thévenot, 1999: 372). In contrast, within this regime of justification the state of unworthiness 'is one in which persons fail, stagnate, and lose out, and in which goods are rejected instead of desired' (Boltanski and Thévenot, 2005: 197). Relevant pieces of evidence submitted support of market justifications only 'qualify' as long as they can be treated as exchangeable goods or services. In the case of maize production, the worth of the crop is evaluated only in terms of price and open market competition. Market arguments for the liberalisation of maize trade might include, for example, claims that the free trade of the crop will boost revenue for NAFTA as a whole, or that it is the cheapest method of providing the product for which there is ample demand. In the case of GM maize, trade in the crop was conceived and defended by NAFTA as an integral part of the agreement intended to foster the 'free circulation' of goods – which is the main point of reference in the construction of the North American market.

The rhetoric used in this form of justification is clearly based on the universal applicability of the laws of neoliberal economics. A good example of this is provided by Major Goodman[37] during his contribution to the above-mentioned international forum on gene flow:

> 'The key question that needs to be faced for Mexico and, by analogy, for other countries that are the centre of origin for the worlds' crop species is whether Mexican farmers and Mexican citizens will be allowed the *choice* to profit from newer technologies and methods' (italics mine).

Choice in the exchange of objects with a clear market value is of fundamental importance within market justifications. Devices that are not directly related to a commercial logic are seen to be illegitimate – unless of course they are put in place to help support commoditisation processes. A case in point are State or international regulatory frameworks that have turned all manner of genes into desirable, saleable, marketable things through intellectual property rights (IPRs), most notably though the intervention of the U.S. government and the WTO.

[37] Crop Science Department (North Carolina State University) and also participating at the forum.

Giving a full account of the issues involved is outside the scope of this paper,[38] but it is clear that the commoditisation of the 'vast interior commons' (Scott, 1998) did boost the expansion of the life sciences (e.g. genomics, bioinformatics, and proteomics) and provided an incentive for industry to expand. To summarise: within a market regime of justification, confidence hinges on the tautological and neo-liberal premise that optimal welfare must, by definition, result from market exchange – in this case, of biotechnology and its food products – eventually managed appropriately under the auspices of the WTO and implemented by the nation state.[39]

8.4 Denouncing GM maize: or how opponents justify their arguments

Alternatives to industrial and market justifications abound. In fact, arguments denouncing the rationalities have always circulated in Mexican society. In the recent past however these arguments have not been very strong. The reason for this is that they have been effectively repressed by undemocratic decision-making. Contrary to Canada and the US, in Mexico no important civil society groups were consulted about the terms in which NAFTA – and hence GM maize imports - should be implemented. Effective lobbying was curtailed by Salinas' autocratic Administration (1988-1994), and it is easy to understand why those who were not consulted were the first to experience the negative externalities of a neoliberal NAFTA run wild. From out of this misfortune, however, came a number of 'hurt' groups. It took these groups a long time to eventually identify, confirm and characterise the negative consequences of free trade, but it can now be safely stated that they have constructed their identity vis-à-vis market coordination through NAFTA, and are now articulating their own demands and requirements in a variety of ways. These groups are increasingly refusing to buy into 'expert advice' or adhere to commercial justifications and instead fight for their right to exercise their 'voice' (Hirschmann, 1970) – that is, they are now returning to the public dispute, expressing themselves eloquently and arguing, suggesting, criticising and participating in a world prematurely robbed from them.

Justifications based on civic equality and solidarity ('civic worth')

One important type of variable that was excluded from NAFTA deliberations was food *sovereignty*. As we have seen above, market considerations emphasised food *security* and stressed market competition. In so doing considerations based on solidarity were auspiciously

[38] For example, since the Bayh-Doyle Act of 1980, patents in the US were broadened to include living organisms such as plants and animals and their constituent parts such as seeds, proteins and, importantly, genes and specific DNA sequences. In the 1990s, and under great pressure from the life-science industry and US government encouragement, the WTO initiated the Trade-Related Intellectual Property Agreement (TRIPs) – an avenue for introducing US patent law on a global scale.

[39] See McAfee (2003) for additional justifications often aired by spokespersons from State and line ministries. See Fitting (2006) for an exhaustive overview (and denunciation) of neo-liberal, market-based arguments that support genetic modification of maize.

assumed to be irrelevant or of little concern. Yet these considerations, even though regarded as marginal by national or international political decision-makers, are crucial from the perspective of those whose livelihoods depend on notions of solidarity. It is no surprise, then, that many of the concerns aired against NAFTA and its neoliberal development proposal centre around issues of equal access to markets, collective welfare, or the protection of civil rights that were (and are) rapidly being dismantled. A parallel concern links the idea of solidarity to notions of equality. This is the case in situations where food sovereignty is coupled to farmers' right of access to land – an important issue in Mexico because of the existing, generalised concern that neoliberalism is betraying the achievements of hard-fought land reform that cost many lives. Various forms of these 'civil' justifications are found in relation to maize production, and equality or solidarity is often the guiding logic behind them.

For example, attempts have been made by various farmers' unions, producers' associations and indigenous groups to connect their struggle against what they see an 'unfair' situation with similar disputes elsewhere – such as the MST[40] in Brazil, Vía Campesina, or the wider concerns of the anti-globalisation movement led by international high-profile figures including José Bové and Naomi Klein. These connections were especially visible before and during the (otherwise failed) Cancún round of the WTO in September 2003. What we see in these initiatives is an attempt by these groups to frame their cause in a language of 'solidarity between struggles'. This is often done during meetings where spokespersons provide testimonies of comparable experiences. This language of solidarity and equality is also evident in local, regional or even national or international-level demonstrations organised against the inequalities and injustices originating in free trade. The high level of participation in these demonstrations (especially in international ones such as the alternative Global Summit in Porto Alegre) is used as proof that their (local) arguments are justified and supported by citizens on a global scale. In the case of GM maize, farmers' and indigenous groups are eager to prove that 'public opinion' is in their favour. They do so by circulating petitions among lists of groups; these petitions propose food policies that do not depend on 'imperialist' pressures, and that are cleverly articulated with arguments against GM maize, notions of unfair market organisation and competition. The composition of these lists is as diverse as possible and the lists are often explicitly presented as silent witnesses of public support to a civic cause.

On other fronts too the importance of collective welfare is apparent. One recurrent issue, for example, is that of biopiracy.[41] Thus one may witness many occasions in which the appropriation of indigenous knowledge and genetic material by biotech firms is denounced. These developments run into direct collision with indigenous notions of intellectual property

[40] *Movemento Sim Terra*, literally the 'movement of the landless'.

[41] 'Bioprospecting' (a more euphemistic term for the same mechanism) in the south of Mexico has indeed raised questions of legal and legitimate forms of access. See Brand and Görg (2003) for an overview and interpretation of some of these projects.

rights, and lead to resentment and conflicts with indigenous communities who claim their rights against those of private plant breeders or corporations. Closely related to this is the issue of farmers' rights with respect to their maize varieties.[42] Here, the discontent of local communities (and of their spokespersons) focuses on the injustice done by legislation that is seen to be closely attached to particular (business) interests, and thus not favourable for public collectivities. In the words of Aldo González, spokesperson of UNOSJO:

> 'We have come to where we are thanks to the free production, reproduction and democratic access to the use of seeds; today this right is being crushed and threatened by the genetic manipulation carried out by transnational companies in their labs' (La Jornada, October 27, 2004).

On the question of farmers' rights, public research institutes are also criticised - as when Pat Mooney (director of ETC) accused CIMMYT's[43] scientists of forgetting that their goal was to help farmers, and of instead making their research subservient to transnational biotech companies (La Jornada, October 25, 2004).[44]

Justifications based on tradition and locality ('domestic worth')

Another important category of arguments concerns justifications in which traditions, locality and local ties to a specific place are highly valued. In evaluations of this type trust, personal relations, community, and proximity are all considered central building blocks of the common good. Arguments related to these topics are often found in the call by farmers and indigenous movements to protect national, regional or local culture and patrimony. Both food sovereignty and maize are instrumental to this.

---

[42] The FAO (1996) defined farmers' rights as "rights arising from the past, present, and future contributions of farmers in conserving, improving, and making available plant genetic resources, particularly those in the centers of origin/diversity". Unfortunately, these rights are not protected under TRIPs or UPOV models (see Wiber, 2006 for wider implications) – nor under Mexican Law.

[43] The Mexico-based *Centro Internacional de Mejoramiento del Maíz y el Trigo* is one of the 17 institutes affiliated under the umbrella of the CGIAR (Consultative Group of International Agricultural Research). It is home to the world's largest maize germplasm collection and delegated by the FAO to conserve this patrimony of humankind.

[44] CIMMYT systematically collects new maize varieties evolving under field conditions, and the possibility that genetically contaminated seed from farmers' fields enters the collection is seen as a real threat. So when CIMMYT adopted a "wait and see" position after the discovery of genetic pollution of maize landraces in Oaxaca and Puebla, the institution was openly 'attacked' by those concerned with agro-ecological diversity and the danger associated with uncontrolled gene flow. Of late, this initial "wait and see" position has been changed in favour of one in which CIMMYT publicly supports (and carries out) research on genetic engineering of maize and wheat. This latter move is congruent with recent shifts in international institutions (from FAO to CGIAR) to complement private sector research in genetic modification, and has further fuelled accusations that these public sector institutes are not serving the general, collective good.

Examples of domestic worth are easy to find. They are generally based on the rejection of non-local authority, and rich testimonials from people all over the country focus on the need to preserve a historical patrimony and way of life, upheld in various ways by existing traditions. A case in point is the joint Oaxaca declaration against GM maize, in March 2004, by farmers, indigenous groups, and civil and environmental organisations:

> 'Here, in this part of the world, maize was born. Our grandfathers raised it, and were raised by it when they forged one of the greater civilisations of history. The oldest house of maize stands on our soils. From here it travelled to other parts of the world. We are maize people. The grain is our brother, it is the foundation of our culture, it is the basis of our here and now. It stands in the middle of our daily lives. It is in our dishes and in a quarter of all products we buy in the shop. It is the heart of rural life, and an important ingredient of the city. We are maize people. And we are so against all odds, always struggling against the dominant winds. Farmer and indigenous knowledge about maize has always been looked down upon, repressed and forgotten. This has led to the extinction of many native maize landraces developed through the patient experimentation of our ancestors... Blind, official policy does not take account of the fact that, for us, maize is more than a cereal. Maize summarises our past, defines our present, and provides the basis for our common future.'

These statements linking maize to a common lineage can be heard over and over again, and they are easily associated with the notion of genetic contamination. For example, during a recent declaration on the 'Genetic Independence of Mexico' that drew together social and farmers' organisations, scientists and environmentalists, Pedro Turuseachi from the Sierra Tarahumara in Chihuahua proclaimed:

> 'Maize is the basis of our life, our culture and our economies. With it, we are born, we grow, we die. Because it's an open pollinating crop, transgenic maize necessarily threatens our traditional varieties. The contamination of our seed is an attack to the heart of Indian communities'.[45]

Pedro is certainly not alone in establishing this relationship between transgenes and the threat to a way of life. Consider the words of Elizabeth Páez, an indigenous woman from Veracruz who attended the same declaration:

> 'Men and women peasants: we have created thousands of varieties of maize of different colour, taste, and size that adapt to all maize fields, be they high, flat, dry, or humid. It has been a creation born out of love and care that we feed and that feeds us. It is the heart of what our ancestors have given us, as well as the main heritage that

[45] www.ecoportal.net/layout/set/print/content/view/full/63042

> we have for our children. But now all is threatened by the greed of a few companies who want to appropriate this treasure in order to steal our seed.'[46]

Implicit in the narratives depicted above is a reference to ancient, local craftsmanship which gives life to local villages and that is in danger of extinction. As these and other, similar stories show, lying just beneath the surface of the account is a rich array of distinct cultures, traditions, and crafts that are highly differentiated across regions (*regionalismo*). These 'domestic' types of evaluative frameworks furthermore accentuate the harmonious, quasi-romantic experience of 'living on the land' - that is, a close connection between people and their place, a relationship mediated by maize. This treasured local place is a valued piece of collective heritage, as well as a location of special meaning for the personal biography of those who live there. This pronounced attachment to place gives rise to notions of tradition, *la patria*, or nationalism, and is often presented as equivalent to, or corresponding to maize and food sovereignty. It does not take much imagination to realise that this type of domestic justifications may also accommodate bellicose tones or outright calls for resistance (the social movement called *en defensa del maíz* being a good example for this).

Justifications based on 'green-ness' and environmentalism ('green worth')

Actions and things are worthy, in this 'green' regime of justification, when they maintain or reflect the principles of environmentalism. This may include notions relating to 'cleanliness', 'non-polluting', 'renewable', 'recyclable', 'sustainable', or 'in harmony with nature'. Justifications based on these notions consider the common good to be inextricably linked to awareness and sensitivity to environmental or ecological issues, protection of wilderness, stewardship of environmental resources, and cultivation of various attachments to nature, the land, or the wild (cf. Lamont and Thévenot, 2000). Often, explicit reference is made to future generations – as in the case of indigenous groups that have been living in a specific region for a long time, and want to preserve their way of life so that their grandchildren can have the same attachment to it. The green order of worth is revealed in distinctly 'ecological' qualifications, such as the *criollo* varieties of maize which are valued as something that is good for the people that grow it, but also for 'nature' as they are important building blocks of special ecosystems not found anywhere else.

At a further level, some environmental justifications depart from the political and moral requirements which are shared by e.g. 'market', 'civic', or 'domestic' orders of worth – where common humanity is the point of reference for the evaluation – and propose an extension of the 'community' of reference to include non-humans. An example of this move from anthropocentrism to eco-centrism can be observed in the deep ecology-like arguments used by certain environmental NGOs that value traditional land-use practices not because of their benefit for humanity but because these practices help sustain agro-biodiversity that is

[46] *ibid.*

worth conserving for its own sake. Closely related to this position are the claims often made by local indigenous associations (in the states of Oaxaca, Chiapas and Guerrero) that it is not so much the uniqueness of native maize that is at stake, but rather the singularity of the entire landscape (wherein maize plays a distinct role) – including of course its inhabitants and the special relation they have to the land. This 'nature/culture' continuum is most evident in arguments that ecosystems in which maize plays a vital role are largely man-made. Hence, as in the notion of 'agri/culture', what is conveyed is the idea that the survival of whole ecosystems depends on the everlasting continuation of traditional practices such as shifting cultivation that are based on indigenous knowledge and culture – a culture that uses but at the same time loves and respects nature. (Wo)man, in this 'ecological' portrayal, has metamorphosed from a 'wrong-doer' to a 'do-gooder': s/he is now the 'noble savage', the guardian of biodiversity! This 'flirting' of indigenous groups with what is perceived as 'indigenous knowledge and culture' is important because of its political appeal in some circles (some local and national left-wing politicians are known to be sensitive to 'native' issues, especially since the Zapatista Revolt in 1994), but also because indigenous groups' special attachment to the land makes them the most legitimate spokespersons for the land and the maize with which it is associated.

8.5 On why the certainties invoked by opponents to GM maize might be misplaced

The arguments about the place of maize (and the people associated with it) in Mexican society discussed in the previous section all involved some type of justification. These justifications illustrated the general value of an argument by strategically linking up with culturally legitimate 'standards' of what is to count as a public good and referring – depending on who was making the argument - to evaluative frameworks based on equality, free market, 'green', industrial, or domestic orders of worth. As I argued these 'grammars' are incommensurable – each one providing the possibility to denounce opposing points of view. This incommensurability does not mean, however, that differences will subsist, or that reconciliation is forever ruled out. Indeed, the literature about public controversies speculates that disputes will converge to closure once a 'best way', an 'economic optimum' or an 'efficient solution' is found (which often means once 'sound' science and technology are thrown in).

But how exactly are 'closure' and 'consensus' reached? Here the literature on the topic is less clear, i.e. normativity is given priority over empirical evidence. The pragmatic approach of Boltanski and Thévenot can offer some clues though. Indeed, one of the main pillars in their model is precisely the possibility of reaching agreement between conflicting orders of worth. Following Boltanski and Thévenot (1999: 369), the prospect of finding a middle ground derives from the empirical observation that 'all persons have, on the same day and in the same social space, to use different devices for assessment, including the reference to different types of worth, when they shift from one situation to another'. To illustrate this point, let us take the case of, for example, a molecular biochemist who on one and the same day may have to design an experimental laboratory test of a new bio-assay, then carry out a discussion with commercial managers about the best technical arguments to be used to sell a new laboratory protocol, then go to a lunch where a *diputado* is invited, and, at the end of the day, take off his tie and attend the union meeting of the University Union's executives. The same individual can later come back home, be taken to task by his wife for being late and, after that, go to a movie made by a newly discovered, inspired, young director.[47] As we see in this little example, agents face a whole range of situations in their daily lives, in which assessments and moral evaluations need to be continuously made. Interestingly, situations close to one another in space and time must be justified according to different principles. Contrary to the sociology of, for example, Bourdieu (1984), Boltanski and Thévenot suggest that agents asked to justify their position are fully aware of what they are doing – and why. Actors are thus not seen as 'cultural dopes' (Garfinkel, 1967) moved about by forces outside their understanding. Rather, actors are seen to be competent and capable of drawing on different and sometimes opposing 'regimes of justification' in the space of a single day. In the chaotic context of everyday life, it is easy to see that not all justifications fit easily into one and only one order of worth – thus giving rise to the possibility (both theoretically and empirically) of compromise (Thévenot *et al.,* 2000). Compromise is thus based on agents' fundamental capacity to critically but also constructively engage with opposing points of view – quite simply because they themselves are acquainted with, and mobilise, the principles of these opposing points of view in their daily lives.

As of the time of writing, however, nothing seems further from reconciliation than the opposing positions partaking in the debate about GM maize. On the contrary, the last years have shown a harsh polarisation between contenders, and constructive negotiations or possible compromise seem far-fetched. This is not because of a lack of understanding of each others' positions (all parties involved are acquainted with their rivals' arguments) but rather because the controversy about GM maize is relatively novel, with the parties involved still very busy deploying their arguments (in fact, polarisation is taking place precisely because more and more arguments - each one based on its specific regime of justification – are circulated by the day).

Yet it would not be surprising if, in the middle to longer term, some rapprochement takes place. This will not be based on 'letting go of ones ideals', but rather through assimilation of one or more different orders of worth by way of scientific argument and technological prowess. And herein lies the danger for the anti-GM movement. As the records show domestic, civic, or environmental concerns are usually reduced, de-politicised and made compatible with industrial or commercial goals and logic. Although this may arguably point to closure or the reaching of compromise, in effect it would mean that alternative logics have

[47] An example of a farmer or an NGO activist 'travelling' from situation to situation and in need of 'applying' different orders of worth according to circumstances can equally well be thought of.

been repressed. How can this be possible? How can seemingly incommensurable orders of worth be joined? To understand the mechanism involved, we need to turn not to what makes 'regimes of justification' different (i.e. the principles of moral and political philosophy invoked) but to a specific feature they have in common: their modernist dependence on certainty about the relationship between ends and means.

It is this means/ends rationality in particular that is important for my argument here. Following Latour (1998), it would indeed be easy to show how each particular 'order of worth' defines a scale of 'rights' and 'wrongs' on the basis of an instrumental rationality. For example, for a proponent of GM maize who frames her argument in terms of a market justification the really important issue, the goal, the *end* is that all market participants become rich. The only way to achieve this, she would assert, is through a set of legitimate means: the free and unhampered circulation of goods (including, of course, GM maize) and services among producers, consumers, and merchants who enter into a relationship with one another as businessmen competing in a globalised market open to all. For her, the final test to determine whether the end has been achieved is to assess whether or not additional wealth has been created. Have economic agents seized the opportunities entailed in GM maize through sanctioned means? If so, then the common good has been served; if not, then surely market imperfections or State failures are to blame, and precious time (and money!) will have been squandered in vain. As this short example shows, we see here a nearly absolute certainty that a set of means will lead to a final end. But this notion of certainty about the connection between ends and means permeates all regimes of justification. Witness for example the green lobby's conviction that only the 'noble savage' will salvage the purity of Mexico's maize landraces, thus helping bring about a desired end: environmental sustainability.

The present situation thus appears to involve a plurality of certainties on which contenders base their justifications and denunciations. Yet the question is: will this plurality prevail, or will it be ultimately assimilated by some 'final certainty'? Will the arguments of the ill-defined movement against GM maize be reduced, domesticated and absorbed within industrial or commercial frameworks? At first glance, the answer seems clear. The movement against GM maize cannot survive as long as it clings to the certainties they seem to be so fond of. I argue that this is so because it would be easy to show that a majority of their certainties can easily be assimilated by proponents of GM maize. The easiest certainties to assimilate/absorb are those that are circulated in denunciations made on the basis of an 'environmental' regime of justification. Indeed, many of the denunciations made from the point of view of the environment can easily be accommodated within the industrial regime, making important means (such as traditional or indigenous practices and local knowledge and culture) quite superfluous for the attainment of environmental sustainability. Thus, for example, the claims and certainties of environmentally-oriented opponents of transgenic maize may rapidly disappear in favour of technologies, equipment, or regulations designed to end or reduce genetic pollution. GURTs[48] immediately come to mind here, as well as risk assessment criteria and test methods (such as PCR[49]) based on the science-based (and thus industrial) justification provided by the concept of 'substantial equivalence'[50]. As Latour (1998: 225) beautifully summarises it: 'After the initial cries of horror at the accounts to be balanced, the costs to be met and the equipment to be installed, it is 'business as usual' for ecology in the 'industrial regime'. Indeed, all the certainties used by the green lobby in launching their criticisms become bleak once the captains of industry can show that potential genetic flow can be managed like any other environmental problem by simply extending the production process and monitoring the 'health' of the environment.

Relatively easy to assimilate (though not uncontested) are certainties involved in the denunciations made on the basis of a 'civil' regime of justification. Stating, as opponents of GM maize do, that collective welfare is the highest goal, and that this end can only be achieved through specific means (e.g. public collectivities, committees, representatives, delegates, or elected officials equipped with forms, decrees, procedures or any sort of legal criteria) does not automatically imply that those in favour of GM maize are pushed aside. After all, those dismissing transgenic maize are only one lobby among many, and their definition of what the highest end (collective welfare, the 'General Good') should be can easily be contested by the 'general will' - that is, democratically elected local, regional or national authorities who claim to speak in the name of all, thereby using the same means as those that are advocated by opponents to biotechnology.

More problematic, but not at all impossible to absorb into an industrial or market logic, are the certainties that are cherished within the 'domestic regime'. The belief that a territory, a heritage, a patrimony or a local tradition can best de defended against the evil nature of a de-territorialised, commercial enterprise through the virtues of anything or anyone reliably attached to the locality (ancestors, local *caudillos*, kinship relations, regionalism) has provided some bargaining power vis-à-vis commercial endeavours. Yet the bizarre alliance between conservatives, nature conservationists and conservationists of folklore under the banner of the 'domestic regime' is evidently not infallible. By attacking transgenes one can be, at the same time, both reactionary and modern. But so can proponents of GM maize, who can likewise assimilate old and new within an industrial regime. Witness, for example, the possibility to allow for the experimentation and eventual cultivation of transgenic

[48] Genetic Use Restriction Technologies. The most widely known of these technologies is the one commonly referred to as 'Terminator Technology' – a method to restrict the use of genetically modified plants by causing seed to be sterile. Proponents argue that using this technology prevents possible escape of transgenes from the source material, thus averting any impact upon biodiversity.

[49] Polymerase Chain Reaction, a technique that allows the detection of transgenic DNA.

[50] See Levidow and Carr (2007) for a fascinating account of the way in which changing definitions of 'substantial equivalence' are used to govern social conflict around GMO's, as well as legitimacy problems of regulatory procedures.

maize under the framework of the LBOGM[51]. According to this framework, GM maize can be authorised in geographical areas that fall outside of the areas of origin of the crop.[52] Notwithstanding the difficulty in establishing these areas, science-based 'zonation' in effect squares the arguments of GMO opponents by countering *regionalismo* with even more *regionalismo* in the form of zonation. But other examples come to mind, as is evident in the regulation and certification of some of the products of locality, patrimony or heritage (e.g. regional *delicatessen* with a geographic indication, *tortillas* sold at *Nuestro Maíz* or *Itanoní* outlets). However praiseworthy these initiatives are, it is not unthinkable that in the foreseeable future they too will have to 'prove' (through technical or scientific means) that they are GM-free[53]. Perhaps recourse to nonbinding international treaties to help protect traditional knowledge and resource rights (such as, among others, CBD's Article 8g or its clearing house mechanism, ICESCR[54], ILO's Convention 169[55], or Unesco's World Heritage Convention) might provide some solace, but evidence suggests that these types of 'soft law' generally do little to nothing in countering commercially-oriented, science-based WTO regulations that are locally sanctioned.[56]

8.6 Creative uncertainties

If the broad social movement against GM maize wishes to escape the fate of being reduced to a commercial or industrial logic, then it had better abandon the central axiom that it has in common with them: the relationship between ends and means. Indeed, in all the different regimes of justification the measure, the main point of reference, is common humanity. The main end of all logics that are used to distinguish between right and wrong is always an end that is good *for* humans, according *to* humans. This common anthropocentric truism makes it possible, in all grammars of worth, to state *with certainty* that everything in creation (whether people or things) can be used merely as a means, and that only man is an end in himself. That is exactly why the pro-GM lobby has (or will soon have) no problem with their opponents' claims: these can easily be accommodated within their logics *because*, in the end, it is all about how to make the world a better place *for humans*. Becoming rich (the market

[51] Ley de Bioseguridad de Organismos Genéticamente Modificados, Mexico's Biosafety Law.

[52] This has, in true 'industrial' fashion, led to a science-based debate about where exactly these areas of origin are located. Some NGO's opposing GM maize are legally implicated in this debate without actually seeing the risk involved: a future, complex mosaic of landscapes where areas set aside for transgenic cultivars will be interspersed with a handful of sanctuaries where 'pure' maize landraces and *teocintes* can be found – a pyrrhic victory indeed!

[53] A forerunner to this possibility is already a reality in the EU where a co-existence regime between GM and non-GM products has been put in place. Ironically, the burden of proof is on GM-free products!

[54] International Covenant on Economic, Social and Cultural Rights.

[55] The International Labour Organisation's Convention Concerning Indigenous Peoples in Independent Countries.

[56] See Posey and Dutfield (1996), however, for a fascinating and encouraging account of instances in which traditional resource rights have been successfully defended.

regime), taking science-based decisions to reach efficiency and productivity (the industrial regime), professing the significance of collective welfare (the civic regime), emphasising the importance of whatever is 'local' (the domestic regime), or searching for environmental friendliness so that future generations can make use of finite resources (the 'green' regime): these ends are all compatible, i.e. reducible to one another as long as it is humans who will benefit. The danger of remaining marginal thus looms large for opponents of GM maize. As long as they adhere to modernist certainties favouring a common humanity, their arguments can be reduced and depoliticised.

Yet there is hope. The originality of the social movement against GM maize lies not so much in the denunciation of market or industrial forms of reasoning, but rather in drawing attention to complex associations between all sorts of beings: beliefs, institutions, politics, regulations, consumers, environments, markets, sciences, traditions, biotechnology and so on - not to mention transgenes. In mobilising a language of certainty, the social movement against GM maize unfortunately risks being reduced. But through their *practices* (which fortunately differ from their discourse), the movement adds something new to the controversy. And what it adds is difficult to ignore, reduce or digest by those in the business of simplifying: the interrelatedness between all humans and non-human participants in the dispute. More accurately: the originality lies in the impossibility to decide – beforehand and without shortcutting due political process – what are ends, what are means, and in what form they may be related to one another. In other words, the great novelty of the movement is their 'ecologising' (Latour, 1998), their practice of deploying *uncertainty* in the relationship between means and ends: in opening the question, that is, of the status of complex imbroglios in which humans and non-humans become progressively entangled without assuming, *a priori*, that only humans can aspire to the kingdom of ends. Regrettably, opponents of GM maize seem to be hesitant about their practices: they believe they are denouncing a market, a neoliberal system, or the science of economics, but it is precisely this self-righteous certainty that marginalises it. As Latour (1998: 230) observes more generally about ecological movements, a better grasping of the meaning of their practices would perhaps permit them to finally attain political maturity – and, one would like to add, clout.

So this is the point I want to make: the anti-GM coalition could circumvent the potential reduction and de-politicisation of their position if they could more resolutely turn to, and exploit, the many uncertainties attached to the GM maize controversy. In practice they are already doing so, but their over-lucid indictment of all that is morally objectionable has hindered, or so I contend, their cause. So what are these practices in which uncertainty is deployed beyond the possibility of reduction? Let us turn to two examples. The first one concerns the field spanned by agricultural policies, peasant livelihoods, and trade liberalisation. When GM maize was first launched proponents (backed by the successive Zedillo and Fox Administrations) presumed they would have it their way - hands down. Soon after the discovery of transgenic contamination of local maize landraces, however, the expectations of the pro-lobby went awry. Before long, the seemingly simple proposition: 'GM maize is good for the economy' turned into a fully-fledged nightmare for its advocates. Complexity was growing by the day as opponents countered – term for term – with their accusations (often based on modernist certainties; see the section above on domestic, civic and environmental justifications). Next to these predictable denunciations those countering transgenes started airing a number of anxieties that turned GM maize into an increasingly uncertain enterprise. What would happen to the livelihoods of millions of campesinos once 'the market' was flooded with cheap, imported, American GM maize? What would happen to the legacy of the Revolution once its ideals – land and liberty – were betrayed? What would happen to lost agricultural knowledge once its owners migrated to the U.S. for good? How could Mexican *campesinos* compete with American farmers if the U.S. was heavily subsidising its agriculture, while the Mexican Administration of the moment was busy dismantling the few subsidy schemes that were left? Why had tariffs not been levied on U.S. maize imports – as NAFTA allowed? Who was benefiting from this? What would happen to the right to food once Mexico became dependent on its Northern neighbour for its main staple crop? How about 'food sovereignty' instead of 'food security'? What about 'democracy'? Why did Mexican farmers never get access to the negotiations that were preliminary to the signing of NAFTA? What if *campesinos* betrayed the corporativist institutions of the PRI, because the PRI did not care about farmers anymore? What if maize was indeed the capillary systems of Mexican culture? These and many more questions arose simultaneously in many corners of the country, and soon matters of fact that were taken for granted turned into hotly debated matters of concern. Who would have expected these developments barely half a decade ago? The ingenuity, the creativity of a broad coalition of farmers, scientists, intellectuals, consumers and left-wing politicians made this possible - precisely because they did not buy into the modernist means-ends rationality so heavily defended by GM maize promoters.

The terrain of scientific 'facts' about maize diversity, transgenes, and genetic complexity provides another good example of the novelty and the creativity with which the anti-GM movement raised a further set of concerns. What is at issue here is that the scientific certainty proponents of bio-industry wish to achieve comes at a cost opponents are not willing to bear since it frames the controversy in too narrow a way. Here too then a new set of questions is opened. What is the meaning of the relationship between means and ends when one is talking about maize in a Vavilov centre of diversity? What if the millenary process of fine-tuning and adaptation between farmers, maize and the environments they help sustain is abruptly severed by transgenes? How fit are transgenic maize varieties? What if genetic diversity erodes as a result of the introgression of transgenes? What if it is not the transgenes themselves that bring about genetic erosion, but rather farmers' and consumers' *perceptions* that landraces could be contaminated, thus possibly causing rejection of these varieties and thus triggering a direct loss of biodiversity (see e.g. Bellon and Berthaud, 2006)? Will Monarch butterflies – which feed on maize pollen – still have a place to live, or will they have to succumb to *Bt*-insecticides in order to satisfy an industry's wishes? What happens if travel-happy transgenes cross over and affect *teocintes*, the wild relatives of maize? How will super weeds that have become resistant to *RoundUp* herbicide be dealt with? And what if assessing the dangers involved (e.g. through risk management processes) is inadequate because the principles upon which these assessments are based (U.S. agriculture) do not apply to Mexican agriculture - where genetic, ecological, economic, social and cultural issues and factors are completely different?[57] Are transgenes as safe as some scientists or the FDA tell us, or do they cause cancer and other degenerative (and possibly hereditary) diseases? What we see in these creative but uncomfortable questions is that novel associations between previously unconnected entities attain – once connected – a force of their own. Providing an 'objective' scientific answer to this type of questions will prove difficult if not outright impossible – not just because scientific evidence about the pros and cons of GMOs is controversial, but also because answers involve fundamental political issues that boil down to the question: are means just that – means – or do means have the right to be seen as ends in themselves?

8.7 Conclusion

In 'stable' situations, different regimes of justification may be alluded to in order to reach closure through compromise. When no compromise can be reached, it is often by means of reduction, de-politicisation and assimilation that one grammar of worth achieves worthiness and imposes itself upon others. But in circumstances where innovations proliferate, when the number of items to be taken into consideration changes continually (many of them invoking concerns that cannot be contained by scientific risk assessment) a new language, a creative rhetoric, novel forms of justification emerge (Callon, 1998). This is particularly so in the virulent disputes surrounding genetically modified organisms (GMOs). A typical example is the controversy over GM maize in Mexico. Since the social movement against transgenic maize does not know how genes, environments, policies or people will behave in gene flow situations, a re-politicisation of the debate has ensued. Indeed, clashes between certainty and uncertainty proliferate in the dispute, and what is at stake is not a mere settling of issues on which there is no previous, generalised consent. Rather, the controversy involves the political and moral question as to who and what may aspire to the kingdom of ends - that is, whether or not previously marginalised farmers, consumers, genes, or ecosystems are something more than merely a means in the achievement of a particular (commercial or industrial) collective good. What we are witnessing is thus the emergence of an entirely different playing field, because whatever is done involves unexpected consequences.

Interestingly, the completely unexpected feature in this controversy is the outright attack on the idea that science, technology or economics will provide the certainties needed (e.g. in the form of efficient end-of-pipe technologies, economic optima, means to ends, etc.). However, as the debate shows this is very problematic in cases with high moral and political content

[57] Cf. Soleri and Cleveland (2006).

- such as genetic modification. Expert knowledge may indeed be an effective instrument to shortcut due political process and keep the crowds at bay, but it does not provide any guidance whatsoever on the issue of how to live one's life, or in what world one wishes to live. Indeed, and so the practices of opponents to transgenic maize show, the problem with proponents' views is that just around the corner of every scientific, technical or economic certainty there is an uncertainty looming large. As the GM maize controversy suggests, such situations are probably the rule and not the exception. Instead of extinguishing moral and political fires science, technology and economics generate uncertainty and even outright ignorance[58]. But ignorance, uncertainty or a lack of consensus do not necessarily lead to paralysis.

On the contrary, as in the case of opposition to GM maize, uncertainty opens up a range of new moral, ethical and political spaces. What I mean to say here is that the politics of the anti-GM movement indeed change by aligning against the marginalising closures of modernist epistemological certainty. While proponents of genetic engineering had certainty and objectivity on their side, those hostile to GM mobilise all the opposites: uncertainty, open-endedness, and the inability to differentiate between subjective and objective. The broad coalition against GM is practically turning these qualities into their guiding principles (i.e. an order of justification), and this allows its participants to bring to the foreground what has been pushed to the background, repressed before. In so doing, they are stirring up politics simply because many important actors (consumers, academics, farmers, and so on) are now entertaining sufficient doubt about the assurances of the pro-GM lobby. Indeed, the neo-liberal tendency to justify the elimination of Mexico's cultural and agro-biological diversity on the basis of the assumption that they are hindering 'development' is slowly - but steadily - making way for alternatives demanding that maize and the peasant way of life be respected and assessed through evaluative frameworks that do not take market values as the only benchmark. In these alternatives, farmers and their beliefs are no longer an anachronism; local maize landraces and the environments they are a part of cannot be seen as part of a natural history museum; genes are no longer entities one can switch on and off at will. Rather, these elements have become virtuous – a collective good worthy of a place of their own.

In contrasting their practices with commercial justifications, the social movement against GM maize shows that farmers, genes and a host of other entities that have become progressively entangled in the controversy attain a new identity once it becomes uncertain whether they are simply means in the schemes of others, or fully-fledged ends in themselves. These more 'ecological' identities contrast sharply with neo-liberal and reductionist narratives and prevent cultural and agro-biodiversity from being treated as contingent 'externalities' of an inevitable process of technical and economic 'development'. Instead, by putting uncertainty on the agenda, collective 'ecological' identities demand renewed attention and political

[58] Ignorance in the sense of situations of radical uncertainty where nothing is known.

space. What has been repressed through simplification (peasants, maize, genes, beliefs and so on) forcefully returns to the (inter)national agenda. By linking their destiny to the principle of uncertainty, the repressed may be able to construct for themselves a new collective identity. In the process they may achieve a complete metamorphosis, and evolve from that which had to be given up to that which has to be cherished and respected on its own terms. The repressed have returned, and they are here to recover their future.

References

Bartra, A., 2004. Rebellious cornfields: towards foods and labour self-sufficiency. In: Otero, G. (ed.) Mexico in transition. Neoliberal globalism, the state and civil society. Zed Books, London, New York, pp. 18-36.

Bauman, Z., 1992. Intimations of postmodernity. Routledge, London.

Bellon, M. and J. Berthaud, 2006. Traditional Mexican agricultural systems and the potential impacts of transgenic varieties on maize diversity. Agriculture and Human Values 23: 3-14.

Boltanski, L. and Thévenot, L., 1991. De la justification: Les économies de la grandeur. Gallimard, Paris.

Boltanski, L. and Thévenot, L., 1999. The sociology of critical capacity. European Journal of Social Theory 2: 359-377.

Boltanski, L. and Thévenot, L., 2005, On justification. The economies of worth. Princeton University Press, Princeton.

Bourdieu, P., 1984. Distinction: A social critique of the judgment of taste. MA: Harvard University Press, Cambridge.

Brand, U. and C. Görg, 2003. The state and the regulation of biodiversity. International biopolitics and the case of Mexico. Geoforum 34: 221-233.

Callon, M., 1998. The Laws of the markets. Blackwell Publishers, The Sociological Review, Oxford, pp. 244-269.

Callon, M. and A. Rip, 1986. Mapping the dynamics of science and technology. MacMillan, London.

Douglas, M., 1966 [1994]. Purity and danger. Routledge, London, New York.

FAO, 1996. Gender, biodiversity, and local knowledge systems (links) for strengthening agricultural and rural development. Project document. Women in development service, February 1996.

Fitting, E., 2006. Importing corn, exporting labor: The neoliberal corn regime, GMOs, and the erosion of Mexican biodiversity. Agriculture and Human Values 23: 15-26.

Garfinkel, H., 1967. Studies in Ethnomethodology. Englewood Cliffs, Prentice-Hall, New York.

Hirschman, A., 1970. Exit, voice, and loyalty: responses to decline in firms, organizations, and states. Harvard University Press, Cambridge.

Knorr-Cetina, K., 1981. The manufacture of knowledge. Pergamon Press, Oxford.

Kuhn, T., 1970. The structure of scientific revolutions. University of Chicago Press, Chicago.

La Jornada, 2004. Available at: http://www.jornada.unam.mx/ultimas/

Lamont, M. and L. Thevenot (eds.), 2000. Rethinking comparative cultural sociology: repertoires of evaluation in France and the United States. Cambridge University Press, Cambridge.

Latour, B., 1987. Science in action. Harvard University Press, Cambridge.
Latour, B., 1998. To modernise or to ecologise? That is the question. In: B. Braun and N. Castree, N. (eds.), Remaking reality. Routledge, London, New York.
Latour, B., 1999. Pandora's hope: Essays in the reality of science studies. Harvard University Press, Cambridge.
Levidow, L. and S. Carr, 2007. Recasting 'substantial equivalence': Transatlantic governance of GM food. Science, Technology & Human Values 32: 26-64.
Lomnitz, L., 1977. Networks and marginality: Life in a Mexican shantytown. Academic Press, New York.
Long, N. and J.D. Van der Ploeg, 1989. Demythologising planned intervention: An actor perspective. Sociologia Ruralis 29: 226-249.
McAfee, K., 2003. Corn culture and dangerous DNA: real and imagined consequences of maize transgene flow in Oaxaca. Journal of Latin American Geography 2: 18-42.
Posey, D.A. and G. Dutfield, 1996. Beyond intellectual property. Towards traditional resource rights for indigenous peoples and local communities. IDRC, Ottawa.
Quist, D. and I. Chapela, 2001. Transgenic DNA introgressed into traditional maize landraces in Oaxaca, Mexico. Nature 414: 541-543.
Schwentesius, R., M.A. Gómez, J.L. Calva and L. Hernández, 2003. El campo aguanta más? Universidad Autónoma de Chapingo, México.
Scott, D., 2003. Science and the consequences of mistruct: Lessons from recent GM controversies. Journal of Agricultural and Environmental Ethics 16: 569-582.
Scott, J.C., 1998. Seeing like a state. Yale University Press, New haven, London.
Serres, M., 1997. The troubadour of knowledge. The University of Michigan Press, Ann Arbor.
Soleri, D. and D.A. Cleveland, 2006. Transgenic maize and Mexican maize diversity: Risky synergy? Agriculture and Human Values 23: 27-31.
Thévenot, L., M. Moody and C. Lafaye, 2000. Forms of valuing nature: arguments and modes of justification in French and American environmental disputes. In: M. Lamont and L. Thévenot (eds.), Rethinking comparative sociology: Repertoires of evaluation in France and the United States. Cambridge University Press, Cambridge, pp. 229-272.
Wiber, M., 2006. Intellectual property rights and food security - The international legal battle over patenting staple crops. Paper prepared for the 25th Anniversary Celebrations of the Law and Governance Group, Wageningen, 14 December 2006, pp. 32.

Part 2: Theoretical reflections

9. Paradigmatic changes

Their significance for collaborative research

Gerard Verschoor

9.1 An uneasy feeling: making a difference

In our different ways – whether intuitively or otherwise - we all would very much like to add our little grain of salt to history. To make a difference. And we all have – again, in our different ways – a more or less clearly defined idea of how we would like to achieve this. For many of us working in the social sciences this often translates into a different attitude towards research. Research not on or about social subjects but for them and, ideally, with them. So that we can create space for change together. But oftentimes we find ourselves running up against a wall, for at the beginning of the 21st century collaborative research between 'us' (the researchers or so-called experts) and 'them' (the researched or so-called lay people) is, unfortunately, the pariah of social science.

So we want to make a difference, for sure, but there is this really big obstacle out there: received wisdom about what is 'good' and what is 'bad' social science – 'bad' and 'unscientific' often being associated with a non- or less-hierarchical collaboration between researchers and researched. In this chapter I broadly describe why this is so. But I also want to argue that things can be different – that things do not necessarily have to stay the way they are. To argue this we need to retrace the path taken by (social) science a bit. To do so, I have set up the chapter as follows. First, via a short thought experiment, I present the main argument. I then provide a short historical account of the way in which Science has produced a status quo where collaborative research is the clear outcast. This is followed by a number of sections that focus on attempts in (social) science to open up spaces for imagining and carrying out alternative, dissenting modes of research such as nineteenth-century Philosophical Romanticism and its heirs: the attack on the 'orthodox consensus', work on ideology and discourse, feminism, poststructuralism and finally the 'turn to practice'.

The idea behind this setup is that it enables one to find the different turns and junctions where our scientific ancestors buried some tools that were meant for us, today. Imaginative tools that might help us carve out a more respectable standing for collaborative research. The title of this chapter conveniently sums up this task. Against this background though I hasten to add that I do not see this chapter as providing recipes for carrying out collaborative research. My task is much more humble: I conceive of it as a source of inspiration and an invitation to seek out the many treasures that lay hidden along the historic path of (social) scientific thought. Most importantly, I would like the reader to finish the chapter asking questions such as: 'what (social) science world have I been sold?', 'in what (social) science world do I want to live?'. It is not my intention then to provide ready-made answers or simplistic solutions to these questions, but I modestly hope to help you find a solution to that most demanding of questions: 'how can I make a difference?'

9.2 The argument: presence/absence

Let me start this section with a simple thought experiment. *When I write 'Science': what comes to mind?* (Please take your time – no need to rush things at this point!)

Let me bet: you associated 'Science' with white coats? Laboratories? Articles? Technology? Knowledge? Universities? Well I guess we all have our ideas about it and, depending on how much time we spent thinking about it, these ideas may range from the very vague to the more concrete. Indeed, in our day 'Science' has become some kind of an 'elevator concept' (Hacking, 1999) that serves to raise the level of a conversation though its meaning has remained remarkably free-floating i.e. not stable or transparent. Yet I asked you to engage in this thought experiment for another purpose: it should make clear that, once you start thinking about 'Science', you simultaneously stop thinking about other things. In other words: when you say (or think) 'Science' you are not saying (or thinking) 'non-Science'. This insight would be all the more obvious if I would ask you to repeat the same experiment in a more concrete way – let us say, by asking you to *write* down an account of what Science is *for you*. After a while you would probably produce a list with the characteristics of Science. Or come up with a definition of what Science *is*. Alternatively you could write a little essay about the way Science works. Or the way it relates to technology for example. But as you produce these lists, definitions, descriptions or accounts you at the same time leave untouched an infinite number of elements that *could* have been part of your story but that somehow you decided not to take on board. So the thing is this: *to make present* (to write, to talk) is, at one and the same time, *to render absent* (to exclude, to make invisible).

In a more general way then, what this little thought experiment must make clear is that 'making present' or 'bringing to the fore' at the same time implies 'making absent' or 'pushing to the back'. Or, to put it in the words of sociologist John Law (2004), what is being present always depends on what is also being made absent. This process - first discovered by linguists – has provided ammunition for those who, for example, make a living out of denouncing the scientific endeavour. For example, many of the accusations about what is written under the banner of 'postmodernism' would run a bit like this: 'making present' means simplifying; simplifying has the effect of taming complexity; taming complexity means obscuring or pushing all that is deemed irrelevant to the fringes. And as things are pushed to the fringes – as they are made absent – violent repression is often exerted.

Now what does this all have to do with collaborative research, you may ask? Well - everything. For in alliance with the postmodernists I argue that the history of Science has been, also, a history of repression. A repression, notably, of types of research where non-scientists could participate. After all (and this could be easily demonstrated) it is not accidental that collaborative research between experts and lay people scores so low in the academic pecking order. Contrary to the pessimistic or at times fatalistic stance of some postmodernists though, I want to point to something else that is entailed in repression, in making absent. Namely that repression (as both psychoanalysis and work in the history of science tells us) may also work *productively*. In psychoanalysis, for example, that which is made absent, which remains hidden - in short all that is repressed or suppressed in the unconscious - may emerge in the form of dreams. Dreams that may disrupt a person's behaviour, for example. The scientific analogy of this is provided by historian of science Kuhn (1970), who describes the way in which accumulated knowledge that has previously been suppressed may 'pop up' and produce scientific revolutions leading to a change of scientific paradigm. Thus in science, as in the unconscious, dissenting forces are produced as a result of the work of simplification and repression. And it is to these forces that I turn in this chapter. Forces that have accumulated through the years and that, if employed creatively, may prove fruitful for those who cherish collaborative research in all its different guises. For people like you and me who are constantly on the lookout for ways to build arguments in favour of more democratic and emancipatory forms of knowing and doing – in short, of carrying out Science.

The forces of dissent I am talking about have accumulated over a long period of time. They have appeared at different points in history, and have been articulated by a variety of people in numerous forms. It is not my intention to fully delve into this history here – first because I am not equipped to do so, but also because such an enterprise would at least necessitate a number of volumes. In what follows, I concentrate only on a small but interesting period (let us say the past 50 years) in which developments in social science have opened up forces that may lead to new possibilities for carrying out alternative types of Science. By turning to the social sciences I do not in any way intend to say that the emerging insights and possibilities might not be relevant to natural scientists. On the contrary, the argument of much of the contemporary critique of social science is that perhaps the segregation between the sciences (one side studying Nature, the other team studying Society) may have been premature, thus calling for a rethink of the relationship between natural and social sciences in the first place. In short, what follows is an invitation to inter-, intra- or multidisciplinary science wherein the role of so-called 'non-scientists' becomes much more important (if not paramount).

9.3 Science as othering: from the Enlightenment (via the Greeks) to modern-day social science

As stated above, digging deep into the history of science and the way it has made invisible other options would be too tall an order. Yet I think it is important to quickly rehearse some of the general features that have accompanied the production of modern science. It is only against this general backcloth that the production of alternatives acquires its sharpness and detail.

Modern-day science – that is, science as we normally think of it – emerged somewhere in the 18th century during a movement called the Enlightenment. The roots of this movement, as historians demonstrate, lay in the rather unique developments that took place between the 13th and the 17th Century when slowly but steadily the character of the world started to change. With the advent of new navigation technologies, European nations quickly expanded into most 'undiscovered' corners of the world. Soon advances in the field of health and technology also boosted industries, the division of labour, population growth, and urbanisation. Indeed, during a relatively short time span the dynamics of change created an incipient 'global village' where new developments and insights succeeded each other at a previously unimaginable speed. These were interesting times, and nothing could no longer be taken for granted. Economies based on reciprocity and exchange gave way to inequalities and market rationalities, simplicity made way for complexity, and cherished certainties made way for uncertainties. All in all, 'Modernity' (the term used to convey the nature of the changes) seemed an ambivalent thing – in the sense that nobody knew whether it was good or bad news.

Trying to make sense of these revolutionary happenings, European intellectuals (often called *philosophes* in the 18th Century) of the likes of Voltaire, Montesquieu, Locke, Newton, Rousseau, Diderot, Hume, Kant or Adam Smith each embarked on his (note that it was never a *her*) version of a project (aptly called The Enlightenment – literally 'throwing light on the facts') that together combined into a tight cluster. Although their set of ideas may not have been very coherent – after all, the *philosophes* were experimenting with new thoughts for the first time, extricating themselves from centuries of obscurantism – they were united by a number of common threads. Minimally, as Hamilton (1992) tells us, they agreed that the events of Modernity could be *known*, and that in order to do so knowledge had to be organised, following Descartes and Pascal, in a *rational* way - meaning by this that knowledge had to be based on clear ideas that were independent of individual experience, and which could be demonstrated to any reasonably thinking person. To this was added the notion of *empiricism* or the idea that all knowledge (about Nature or about Society) should be based upon empirical *facts* that is, things that every person can experience through their senses. The *philosophes* furthermore agreed that it was through the *scientific method* devised by Descartes (and that could be used everywhere because the method was seen to produce laws that applied in the entire universe) that *progress* could be achieved – in the sense that the natural and social condition of human beings (their well-being, their level of happiness) could be improved by the application of science and reason.

So by the late 18th century there was broad consensus that Science was *the* way for understanding modern developments on the one hand, and for improving the human condition on the other. What was less clear though was *who exactly* could partake in the exercise of science. Remember that in the 18th century there were no laboratories or all those things you associated with Science in the little thought experiment above, so it was not clear who exactly could claim the status of 'scientist'. Heated debates were carried out on the issue, with some *philosophes* inclined to put science in the hands of intellectuals while others argued that the scientific enterprise was a universal human concern and therefore open to all 'reasonable' persons. In fact, one of the greatest works of The Enlightenment – Diderot's *Encyclópedie* - argued for collaboration between theoreticians and workers, and for the respect of the artisan's 'flair', knack, and intuition (Stengers, 1997: 117). However, a wrong turn was taken to end this dispute when some of the *philosophes* turned to the ancient Greeks as a source of inspiration and thereby reintroduced the supreme position of the expert. There and then, a modern-day asymmetry between the expert and his other, the layperson, was created.

For the purpose of my argument it is important to take a brief look at the distinction the ancient Greeks made between 'legitimate' knowledge held by experts (the carriers of Reason) and the 'beliefs' of ordinary people. The divide can be traced back to ancient Athens where a few patricians and scientists afraid of 'mob rule' (the fear that if reason does not rule, then mere force will take over) sought to turn the balance of forces to their advantage. The answer, as we know from the writings of Plato and Socrates, was found in the argument that the 'power of Reason' - which was thought to rule 'over Gods and Men' (Serres 1997) – and not 'lowly passions and interests' should be used to reach closure in times of controversy and debate. Importantly, scientists (and patricians) claimed that they were in an advantageous position because their empirical studies (in the case of scientists) or their social status (in the case of patricians) gave them direct access to Reason. Twenty-five hundred years ago then, the happy few already mobilised 'Science' and 'Reason' to justify their natural superiority in trying to silence the voice of 'ten thousand fools'. The role of laypersons in the creation of knowledge was, from the very beginning, thwarted by the messengers of Reason who feared that access to the 'Reality' or 'Truth' would be seized by whatever the crowd (moved by lowly passions and impure interests) felt was real or right (cf. Latour, 1999). What I am saying here is – no more and no less - that the marginalisation of laypersons in controversies about public issues has been a *political* machination all along.

Space does not allow delving into the way in which this machination took effect, but the fact is that two-thousand years after the Greeks put in place a divide between experts and lay people not much had changed really. Of course, Science (with a capital S) experienced a lengthy absence from centre stage as during the Middle Ages the idea reigned that only through God could one establish a relationship with the world. The Scholastic method, the protestant Reformation, Humanism and, finally the Renaissance however saw to it that Science was put in charge once again. By the 18th century Enlightenment Science and its obligatory travel mate (the divide between experts and ordinary citizens) were undisputable features of the intellectual landscape.

The Enlightenment's way of looking at things did not go uncontested though. By the beginning of the 19th Century a parallel intellectual movement opposing the Enlightenment's version of Modernity was very much on the scene. This movement, often referred to as philosophical Romanticism, had its roots in Germany and was basically a cultural revolution initiated by the German bourgeoisie who were afraid to be overtaken by the French Revolution and its tenets. Let us remember that, at that time, Germany was made up of many principalities with a known fighting tradition against everything French. As a result, the growing German middle class (with strong nationalistic elements) turned away from French ideas and solutions in their quest for a more modern society. Indeed, they referred with horror to the 'bourgeois revolution' that had taken place in France, and abhorred ideas of citizenship as professed by their western neighbours. In fact, the Romanticists' reading of the French Revolution was that it had everything to do with the Enlightenment's ideals. Hence the Enlightenment had to be resisted.

As a result, they challenged the Enlightenment by positing that rational scientific knowledge (especially of society) could not be reliably acquired and used to improve society. Rather, Romanticists held that 'knowledge' was mostly a personal affair that was largely intuitive and imbued with 'meanings' that were context-dependent. Because of this, Romanticists were in fact claiming that there existed different truths in different places. Therefore, they argued that scientific knowledge could not easily be generalised and thus had limited applicability. At most, they claimed, science could only provide partial vistas. As concerns the notion of emancipation (a concept central to the Enlightenment) Romanticists claimed that it could never be reached through science, but only through personal experience or – why not? – collective learning by the putting together of individual experiences.

The modernists (those pursuing the Enlightenment's ideals) of course did not keep quiet in the face of the Romanticists' attacks. In fact, they did everything in their power to thwart the movement. It falls outside the scope of this Chapter to outline the way in which this was achieved.[59] The end result of the Modernists' efforts though is that, today, there is a (Euro-American) consensus about how research (and more generally Science) should be carried out. The backbone of this generalised agreement (or 'orthodox consensus' as it is often referred to) is that Science should be protected from distortions (like those introduced by Romanticists). The idea of Science having to be 'protected' is closely related to 'scientism'. More specifically, it is linked to 'empiricism' and to 'positivism'. Scientism, empiricism and positivism are alive and kicking to our day. Look around you and you will quickly discover it: it is advanced by all those who appeal to reason and rationality as a way to keep at bay the masses who would also like to have a say in deciding what is right and wrong, what is good and bad – as in controversies surrounding genetically modified organisms, global warming, mad cow disease or biofuels. In these controversies, it is always the same: reason should prevail. And by 'Reason' (capital R) is always meant reason of the scientistic, empiricist, and

[59] Good overviews of the institutionalization of 'modernist' science are given by authors working in the realm of what is called the 'social studies of science' or in political philosophy. In a nutshell, what they argue is that modernism is a continuous battle, a struggle to purge scientific institutions of their 'impurities' (i.e. their *not* conforming to the canons of a positivist epistemology à la Popper).

positivist sort which should be disentangled from the 'passions and the interests' of non-scientific actors, and instead be associated solely with 'logic, with facts, and with methods for determining the facts' (Merton, 1973). It should be clear by now who are meant to be able to carry out this task...

As this necessarily short and caricaturesque portrayal of the history of Science has hopefully made clear, the reason why it is extremely difficult to change the academic pecking order – usually going from 'hard', 'natural' and quantitative to 'soft', 'social' and qualitative, and from 'theoretical' to 'practical' approaches - is that a Great Divide has been constructed by the experts with the purpose of keeping the non-experts at bay. But this Great Divide is more than that (a Divide): the way it works is that one side (scientism, empiricism, positivism) 'eclipses' or tends to erase its Other: all that does not conform to 'Reason'. Yet this repressed Other is difficult to contain and, as I argued in the introduction, has an inclination to resurface - as was indeed the case during Romanticism. Although Romanticism did not succeed in imposing itself in the scientific establishment, there are no grounds for believing that all that is pushed to the fringes by modernism should remain there forever. In fact, from the middle to the late twentieth century a number of developments in (social) science – they could aptly be called the heirs to Romanticism - again nagged at the philosophical groundings of modernist (scientistic, empiricist, positivist) science. In the next sections I summarise these developments, and point to the possibilities they open up for collaborative research. In so doing, I hope to carve out some space for reconfiguring the asymmetrical balance of forces between the happy, knowledgeable few and those who have been silenced for undue political reasons.

9.4 Giving a face to the invisible: developments in contemporary social theory

The main backdrop of classical social theorists' motivation was the Enlightenment and its idea that sociology was a 'natural science'. In other words, the Enlightenment's thinkers believed that the world could be viewed in a scientific way – that is, that there were well-founded ways (rationalism, empiricism, the positivist method) to know about the world, and these ways were better than the alternatives. The Enlightenment's project also, however, suggested that 'good', 'scientific' knowledge is knowledge that is linked to prediction (and in many cases control). This idea is closely linked to the belief that 'the good life' may be planned and implemented so long as the knowledge on which planning and implementation are based is 'sound' (that is, based on scientific results). One can find this commitment to emancipation throughout the history of social theory – from Marx through Durkheim to Parsons (cf. Bauman, 1989).

In the wake of these ideas modern social theory has been associated with a conviction to provide a 'scientific', grand, and all-encompassing view of modern society with a view to plan, control or otherwise improve the human condition. This conviction endures well into the present, not least due to the work of 'the classics': Marx, Durkheim, and Weber. Marx for instance equated Modernity with capitalism. He believed that the rise and dynamics of this system could be explained in terms of economic and social contradictions and class conflict, both of which could (crucially!) be 'uncovered' through the use of scientific method. The same attitude towards method was evident in Weber, who saw Modernity as a process of rationalisation which could in turn be explicated through rigorous study of the interplay between materials and meanings. Durkheim finally was a fiery advocate of positivist, empiricist science which he applied to demonstrate the link between Modernity and the division of labour.

Again, what these scientists have in common is the certainty that 'their' science (rational, empiricist, and positivist as it was) held the key to understanding the changes that accompanied Modernity. Yet during the first sixty years of the twentieth century Modernity intensified at an unanticipated speed. Rapid social change, globalisation, the commoditisation of information, and the advent of 'scientific' horrors (the Holocaust, the atomic bomb, chemical pollution of just about everything) in fact left many in doubt as to whether science was the solution to these problems, or its root cause. Indeed, general uncertainty and scepticism about the possibility of progress loomed large by the 1960s and so it was unavoidable that, as authors such as Giddens and Beck argue, people searched for meaning by reflexively examining and questioning the significance of their actions. However, they argue, this also applies for social theory. Social theory after the 1960s thus tended to examine and question its own roots (cf. Giddens, 1990; Beck *et al.*, 1994). As a result, the status of (rational, positivist) science was under attack by hordes of 'disenchanted' intellectuals, finally leading to the erosion of social science's Grand Narratives (i.e. the stories told about society by Marx, Weber, Durkheim and their successors that seemed to encompass everything). I now turn to some of the reflexive, paradigmatic shifts

Meaning and (in)determinism

The first shots to be fired at the Grand Narratives of social science came from an unlikely quarter: Garfinkel's (1967) programme on ethnomethodology. Taking a cue from the Romanticists, Garfinkel's techniques set out to 'expose' the fragility of social reality (which he saw as being constructed 'on the spot' through conversation in micro-encounters) and its inherent negotiability. For Garfinkel this permanently under-determined notion of social reality totally revolutionised the role of the social scientist. Instead of uncovering the *commonalities* of *the* social system (its order, its structured spaces for interaction, and the pre-determined behaviour of actors as an effect of mechanisms of domination or shared values) Garfinkel proposed that the social scientist focus on the explanation of *differences*. For him, social actors were no 'cultural dopes' moved by invisible strings held by some 'invisible hand' but rather *agents* who could achieve order in a variety of situations and in a variety of spaces. The scientific establishment hammered Garfinkel badly after it took notice of his programme. But ethnomethodology took root quickly and soon adopted Alfred Schutz (the 'father' of the idea of 'multiple realities') as its spiritual ancestor, turned to Ludwig Wittgenstein to borrow the notion of 'language games', and sought academic respectability in the writings of Hans-Georg Gadamer and his dismissal of universal and objective Truth.

The movement initiated by Garfinkel and the ethnomethodologists resonated with other developments in the social sciences of the time. Some of them (most notably symbolic interactionism) referred to the question of where to locate meaning. Various answers to this question existed, and an important one (Weber) suggested that, although meanings may differ in kind, they may nonetheless be brought within a causal control idiom (and thus made to fit positivist science). But there were other answers to the question. These answers (most poignantly made by symbolic interactionists, cf. Blumer, 1969) drew on the work of early twentieth-century theorists such as Georg Simmel, Walter Benjamin and George Herbert Mead. The gist of these thinkers was that meaning was not made up of a set of discoverable processes, systems or structures that could be 'discovered' in the sense that scientists could draw the curtains on some pre-existing reality waiting to be reported on. Instead, their argument was that meaning (or better: meanings, in the plural) was too complex for such a task. Hence their conviction that as far as social science is concerned, indeterminism rather than determinism reigns. At best, they posited, theories about meaning could only catch a glimpse of it. Hence - since all views (theories!) are partial – there are several ways of reporting on meaning: there is no legitimacy anymore for a 'God's eye view' that can encapsulate everything all at once.

Science and the critique of ideology

A parallel development that has helped speed up the dissolution of Enlightenment thinking is that which occurred during the late 60s and early 70s of the last century on ideology (cf. Althusser, 1971). The argument (which leans heavily on Marxist analysis) runs something like this. There are two major classes in history: the bourgeoisie or ruling class and the proletariat or working class. The main interest of the first is to sustain (and therefore conceal) exploitation, while the primary interest of the second is geared towards challenging (and therefore revealing) exploitation. In Marxist analysis the proletariat could in principle understand *real* class relations – if only because it is in its interest to do so. Hence Marxism (as the spokesperson for the proletariat) was the only science that could access Truth. Bourgeois science in contrast was unable to access Truth precisely because its interests lay in preventing the understanding of class relations. Bourgeois science thus produced 'false knowledge' or, in Marxist terminology, 'ideology'.

In practical terms though this presented a number of problems, the most evident one being the question of why the majority of workers (the proletariat) believed ideology (or 'false', 'ideological' stories) instead of seeing the truth of Marxism. One of the responses (a negative, critical one) was that that this was due to 'propaganda' which deluded and misinformed workers about Truth. A more positive response maintained that, although ideologies were false, they were also and at the same time practically workable tools for making sense of the world. This latter response thus held that ideologies are stories one learns (because they come from our social context) and believes (because they *work*). By implication however this latter, more positive version of ideology also turned against positivist notions of Marxism, for if ideologies were 'stories' that worked then perhaps Marxism itself was only one more of these stories (in the sense that Marxism is just another practical tool for making sense of the world which works because it comes from, and fits, certain social contexts). Hence Marxist ideas (scientific Truth) were as equally context–based as non-Marxist ideas (false ideologies).

Work on ideology thus argued that theories (including Marxism) are cultural tools that simplify the social and the natural in order to allow us to make sense of the world. In so doing theories assume and presuppose quite a lot. But many of these assumptions and presuppositions – or so the argument ran - are hidden. Thus a 'paradigm' (in Thomas Kuhn's sense) contains explicit claims, but also a lot of (unavoidable) intellectual baggage. And this baggage may be thrown overboard when circumstances change. In other words, theories are never 'neutral'; they organise facts against a setting of social values. Extending this logic it was only a small step to reach the next, important, insight: that there were no grounds to claim a distinction between Truth and falsity, i.e. all knowledges were deemed to be children of their time, place and social context, and hence shared an equal general status[60] (with all that could be said being that some knowledges support domination and conceal unjust distributions, while others do not).

Pierre Bourdieu had a different take on this issue in his variant of the critique of ideology which he called critical sociology. He and his followers shared the diagnosis of earlier work on ideology – the idea that not only theory, but everything around us (i.e. all cultural objects) comes pre-constructed, that is, laden with prejudices and presuppositions. Importantly however, Bourdieu (he was a politically engaged intellectual) argued that it was the task of critical social science to denounce the practices (and their justifications) that led hidden forms of oppression and repression in order to overcome them. According to Bourdieu and Wacquant (1992), social scientific method can be both critical and positive because of the potential awareness it brings with it: that the break with all social pre-constructions is the only adequate way to re-construct – that is, to build a world in a non-oppressive way. This position, inevitably, has serious consequences for the practice of doing research (and reporting on it).

Science as gendered

Just about the same time as work on ideology was being carried out, a strong critique of the maleness of method in the social sciences emerged. This involved affirmative action by

[60] This is what is often referred to as the 'problem of relativism' or the idea that there are no legitimate means for deciding between 'true' and 'false' or between 'facts' and 'ideas'.

women against the small number of women faculty (university staff and professors). As a consequence science was largely being written by men (to the exclusion or Othering of women) who invariably led to the invisibilisation of women's work and of gender relations. This was also a time of trenchant critiques of patriarchal assumptions about gender which permeated all sciences. So, for example, women were seen as less social (Durkheim) or they were thought to carry out their activities in social domains other than those of men (home versus the public sphere, as in American functionalism of the 1950s). As a result different strands of feminism ensued, most notably feminist empiricism, a position referred to as standpoint epistemology (e.g. Harding, 1986; Smith, 1987), and post-modern feminism (e.g. Hekman, 1990).

Feminist empiricism held that social science was sexist and androcentric, and that this was due to the fact that science was not scientific enough. To make it 'scientific', women had to be allowed into social theory in order to make it less patriarchal and thus less biased. Although this position seems to reaffirm the Enlightenment project in the sense of a possibility of finding out about Truth, this strand of feminism also undermined Modernist thought: if social theory was shaped, i.e. improved by the inflow of women, then how could it be called more 'scientific' considering the fact that Science (with a capital S) was not supposed to be related or influenced by its social context?

This undermining of Modernist thinking was given one more push by standpoint epistemology. In a nutshell, what this position entails is that, in analogy with Marxism, there is patriarchal ideology made up of ideas and beliefs which conceal exploitation (this time not of the proletariat by the bourgeoisie but of women by men). This ideology was seen to distort the patriarchal reality of everyday life but (as in the case of the more positive critique of ideology) it was also said to *work* as a tool for women - in some measure, at least. Social science, standpoint epistemologists argued, was a part of patriarchal ideology. It thus produced distorted or biased knowledge. They countered that Truth was only visible to women as an exploited group and that therefore only feminist social science (carried out by women) was truly scientific. Again, we see here an analogy with the Marxist critique of ideology. But as with Marxism, here of course a number of problems arose. Was there only *one* women's viewpoint? And how about the fact that women may have different ethnic backgrounds, be of different ages, belong to different classes, or have different sexual preferences? In the face of these differences, the question to ask was, of course, how to know which one was true.

This in principle relativistic idea that there existed multiple, local women's knowledges led to a further erosion of Enlightenment thinking and was more fully developed by post-modern feminists who were very sceptical about universal, Grand Narratives - including those about women. Their argument was that knowledges/truths were local accomplishments that arose from struggle.[61] To take account of the question of differences - the Achilles heel of standpoint epistemologists - post-modern feminists proposed that in fact one could not talk in terms of 'essential women' but, rather, that women (and men) had fractured identities related to their ethnicity, sexuality, or class position. Interestingly in the context of this book, post-modern feminists seek forms of knowledge/truth which try to achieve links between fractured identities but which also *interfere* with social injustices in order to effect concrete changes.

Perhaps the greatest lesson from the critique of ideology and feminism is the insight that Marxism (or feminism) is shaped by social interests. This claim sometimes goes by the name of relativism or the assertion that all knowledge is contextual. The crucial question here is: does this matter? Some would argue that it does: we need some firm ground for our knowledge, and hence they propose that one finds a way out of relativism. But others say 'No, that's how it is, and we have to come to terms with the fact that all knowledge is a local and practical tool for making sense of the world'. The ramifications of this latter position are important, for it implies the dilution (if not dissolution) of the Enlightenment project and its desire for emancipation.

Science as discourse

The paradigmatic shift towards the idea that science (like all knowledge) is shaped i.e. distorted by its social context, seems to go a long way to debunking the main tenets of the positivist, rationalist and empiricist stance that so characterises Modernist thought. But matters are more complicated still. There is a line of argument in social theory which says that social relations are ordered discursively or, more defiantly still, that everything (including science) is discourse. As a summary approximation, the term represents a logic of social relations which defines the world around us. For example, it defines subjectivity (or what we think we are). But discourse also circumscribes the character of social relations and the power asymmetries and exclusions that go with them. It also defines what counts as truth, as well as the character of the objects that make up the world. Furthermore, discourse is said to be performed or enacted in or by social relations (in this sense, and since there is nothing that falls outside of social relations, everything is said to be 'discourse'). Importantly though, since discourse is constructed, performed or enacted locally (i.e. it travels through networks of a greater or lesser scope) there is the possibility (in analogy to ideology) of the existence of a plurality of discourses that may coexist – even though some discourses are dominant.

The argument draws on semiotics: a linguistic theory that says that the terms in a language are significant not because they relate to the objects which they claim to describe, but rather because of their relationships with one another, like in 'mother/father', 'nature/culture', 'black/white', and so on (think, perhaps, of a *network* of terms related to one another in some

[61] This argument is strongly linked to developments in discourse analysis, see next section.

kind of grammar). Within semiotics, meaning is arbitrary and always depends on what is being associated with what, whom with what, or whom with whom. Like words that can be used to construct different sentences, so meanings may vary in time and space. In this sense, as poststructuralists such as Deleuze, Derrida, Foucault or Lacan would say that there is no single deep structure or grammar that governs action or language. Rather, in every era there are a multitude of grammars, which means that one can identify these discourses as well as the way in which they are created and deployed. The argument then is that everything social (people, social interactions, texts, architectures) can be seen as relational effects, ordered in terms of a discursive syntax.

We (and our theories) are thus syntactical products, and social theory only serves to reproduce the discourses in which we find ourselves locked (i.e. we are 'made in discourse'). This represents a further dissolution of the Enlightenment project: we don't stand outside and describe what we study. Rather, we help reproduce it by forming part of it – and reproducing the power relations that are implied by it, as in the self-policing person (cf. Foucault, 1979). Nearly needless to say, reflexive use of this principle holds great promise for alternative, collaborative research methods that aim to contest practices and narratives of domination.

Practice, symmetry, and multiplicity

The positions discussed so far can be fitted within a general, quasi-historical aim of social theory: that of understanding (and dealing with) the consequences of Modernity. Lately however, we have witnessed the arrival of positions that are difficult to frame within this general history. One of these is advocated by Bruno Latour (e.g. Latour, 1993) whose main aim is to redefine what is commonly understood by social science. This is a difficult task, he argues, because the contemporary meanings of society and science bear little relation to what the founders of social science had in mind when they invented their disciplines: neither one has remained stable enough to deliver the promises of a strong socio-logy (Latour, 2005). His approach (commonly referred to as actor-network theory) thus engages with a number of tricky questions: What is a society? What does the word 'social' mean? What is science? The answers given to these questions by actor-network theorists (e.g. Callon, 1998; Law, 2004; Latour, 1999) vary, but together they form a sufficiently coherent body of work to have set in motion something of a 'paradigm change' in social theory.

As a summary approximation, actor-network theory is based on a number of premises. The first one is that the object of social science is that of social *practice* and *action* (rather than structures or discourses), for example the actions of scientists. In doing so actor-network theory treat all actions as equals (including that of making theories, labelling others, justifying oneself or denouncing others, etc.) without presupposing that certain actions drive, or are driven by others. This refusal to give primacy to accepted ontological categories means that actor-network theory sees as its task the description of the *diversity* of actions it encounters in fieldwork situations. This *multiplicity* of actions is connected to a second premise: that everything in the world is constructed or (perhaps a better term) *enacted*. By this is meant that realities (in the plural) and representations of realities are 'brought into being in a continuous process of production and reproduction, and have no status, standing or reality outside those processes' (Law, 2004: 159).

In the context of this book this insight is crucial because it means that the sciences are constantly making and remaking the realities we inhabit (in the sense that economists, through their methods, produce what is called 'the economy', geographers 'geography' or feminists 'gender awareness'). And here a further premise is thrown in: it's true that sciences make realities, but at the same time the sciences set the *rules* for crafting these realities. And these rules often establish, *a priori*, what is a right and what is a wrong view of doing science – a right way linked to good method and a wrong one tied to bad method. With (you may already have guessed it) 'bad' associated with 'pollution' of method through interference by non-scientists, laypersons, activists and so on... In light of this situation actor-network theorists have asked a pertinent question: why not just do away with the asymmetry and consider *all* forms or practices of creating realities as possibly appropriate methods? (Including, of course, scientific practice as normally understood). This form of unorthodox critique can go quite far. Witness, for example, a provocative Latour (1999: 17-18):

> 'Scientists always stomp around meetings talking about "bridging the two-culture gap" (i.e. the gap between Science and Society), but when scores of people from outside the sciences begin to build just that bridge, they recoil in horror and want to impose the strangest of all gags on free speech since Socrates: only scientists should speak about science!... Just imagine if that slogan were generalised: only politicians should speak about politics, businessmen about business; or even worse: only rats will speak about rats, frogs about frogs, electrons about electrons.'

A closely related body of work is that developed in France by Boltanski and Thévenot in their theory on justification (Boltanski and Thévenot, 1991). Both authors were among the first students of Pierre Bourdieu, but they gradually dissociated themselves from his tradition of critical sociology, arguing that no type of social metaphysics can be validated (or proven false) from any position external to itself. Rather, they posit, there is a plurality of such metaphysics (or 'orders of *grandeur*' as they call it), some of which are incompatible and some of which may arrive at compromise. From their perspective, social scientists have the task of constructing or describing systems of coherent actions which can account for the diversity apparent in society. However, since the description of the social scientist is only one more addition (or re-presentation) to these coherent actions plurality is further enlarged. A key point in Boltanski and Thévenot's work is that they grant this capacity of creating plurality to all actors – only some of which are called 'scientists'. In other words, what they are saying in their critique of critical sociology is not that sociology can no longer be critical, but rather that the *critique* is no longer limited to sociology (or social science).

In this sense, and contrary to, for example, Bourdieu, what these authors call a 'sociology of critical capacity' (Boltanski and Thévenot, 1999) takes seriously the competence of all people to justify their own actions, i.e. denounce the actions of others – thus opening the door to science for the evaluative repertoires of lay actors.

9.5 Conclusion

For some people, the only identity that the sciences can have is that of 'pure' reason. For them science is 'good' only when it is not 'polluted' by politics, passions, or subjectivity. Theirs is 'a desire for mastery, of manipulative, even violent, reason, of the *a priori* negation of anything that cannot be subject to calculation and to the articulation between general law and particular case' (Stengers, 1997: 110). They are the stronger inheritors of the Enlightenment. For others (not necessarily working at universities or research institutes!) science (or technology, objectivity) is seen as a 'danger' to human rights, morality, or subjectivity. We in this book fight against both purges, and maintain that the more a science is connected to non-science, the better the science in question. Bridging the 'two culture gaps' in this way places those of us who argue for collaborative research in a difficult position: we are seen as traitors to both scientism *and* humanism. Wishing away the animosity towards our enterprise will do no good, so we must face the challenge and stand up to it. This chapter tries to provide some ammunition to do so by pointing to interesting options the Enlightenment project repressed but which have now returned in a plurality of ways. As I have shown, work on meaning and (in)determinism, on the critique of ideology, on feminism, on discourse, or on practice, symmetry, and multiplicity has brought these repressed others back onto the agenda of (social) science.

But there is a problem here. Although the paradigmatic shifts sketched out above have led to important changes (above all in the status of rational, empiricist and positivist science) one of the main strongholds of science – its incredible privilege of defining its object, and being in command of the questions to be asked of it - has been left relatively unscathed. Ethnomethodology, critical theory, feminism, poststructuralism, or actor-network theory have all obtained their legitimate, academic credentials. They have earned the right to cohabitate the campuses wherein 'hard core' positivist science reigned unchallenged barely 50 years ago. Yet the problem is that these new academic bedfellows are just that: extensions of a political project geared to keep non-scientists outside the walls of academia. Some exceptions aside, none of the new academic fields takes lay persons in earnest – let alone sees them as co-researchers. Yet if the objective is to help give a face to the voices that were silenced in the name of modernist social science, then clearly the job accomplished so far is still far from complete. This is a story that is still in the making, but I think that in making it some of the paradigmatic shifts mentioned above can nevertheless serve as a source of inspiration. For even though the texts of some authors presented above are often abstract and philosophically demanding, together their work suggests that Truth is no longer the yard stick by which all can be judged. Indeed (and this is probably the most important insight from the paradigmatic changes presented above) there are more arbiters, more 'goods' to be taken into account.

References

Althusser, L., 1971. Ideology and the ideological state apparatuses. In: Althusser, L.(ed.) Lenin and philosophy and other essays. New Left Books, London, pp. 121-173.
Bauman, Z., 1989. Modernity and the Holocaust. Basil Blackwell, Oxford.
Beck, U., A. Giddens and S. Lash, 1994. Reflexive modernization: Politics, tradition and aesthetics in the modern social order. Stanford University Press, Stanford.
Blumer, H., 1969. Sociological implications of the thought of George Herbert Mead. Symbolic Interactionism, Prentice Hall, Englewood, pp. 61-77.
Boltanski, L. and L. Thévenot, 1991. De la justification. Gallimard, Paris.
Boltanski, L. and L. Thévenot, 1999. The sociology of critical capacity. European Journal of Social Theory 2: 359-377.
Bourdieu, P. and L. Wacquant, 1992. An invitation to reflexive sociology. University of Chicago Press, Chicago.
Callon, M., 1998. The laws of the markets. Blackwell Publishers, Oxford.
Foucault, M., 1979. Discipline and punish. The birth of the prison. Vintage Books, New York.
Garfinkel, H., 1967. Studies in ethnomethodology. Prentice-Hall, Englewood Cliffs.
Giddens, A., 1990. The reflexivity of modernity. In: The consequences of modernity. Polity Press, Cambridge, pp. 36-45.
Hacking, I., 1999. The social construction of what? Harvard University Press, Harvard, USA.
Hamilton, P., 1992. The Enlightenment and the birth of social science. In: Formations of modernity. Open University Press, Milton Keynes, pp. 18-69.
Harding, S., 1986. The science question in feminism. Open University Press, Milton Keynes.
Hekman, S., 1990. Gender and knowledge: Elements of a postmodern feminism. Polity Press, Cambridge.
Kuhn, T.S., 1970. Logic of discovery or psychology of research? and Reflections on my critics. In: Lakatos, I. and Musgrave, A., Criticism and the growth of knowledge. Cambridge University Press, Cambridge.
Latour, B., 1993. We have never been modern. Harvard University Press, Cambridge.
Latour, B., 1999. Pandora's hope. Essays on the reality of science studies. Harvard University Press, Cambridge.
Latour, B., 2005. Reassembling the social: An introduction to actor-network theory. Oxford University Press, Oxford.
Law, J., 2004. After method: Mess in social science research. Routledge, London.
Merton, R.K., 1973. The normative structure of science. In: Storer, N.W. (ed.), The sociology of science. Chicago University Press, Chicago, pp. 267-278.
Serres, M., 1997. The troubadour of knowledge. The University of Michigan Press, Ann Arbor.
Smith, D., 1987. The everyday world as problematic. Open University, Milton Keynes, pp. 49-88.
Stengers, I., 1997. Power and invention. University of Minnesota Press, Minneapolis, London.

10. Building bridges to the future

Connecting science to society

Coyan Tromp

10.1 Introduction

Though we all live in a knowledge society nowadays, many countries seem to be faced with a knowledge paradox: while the potential for innovation and social change is there, the transmission of knowledge to society is far from optimal. The current paradigm directing science and research is predominantly focusing on our cognitive-instrumental interests in knowledge, while neglecting our more practical, moral and existential needs. To come up with a science and research model that can do justice to the whole range of needs, we need to found it on a broader conception of rationality than the narrow-angled one that is leading the way now. Only then can we hope to bridge the gap between science and society.

10.2 Changing ideas about the role of science

Current challenges, suitable solutions

In our global society at the beginning of the 21st century, our hunger for knowledge in a way seems greater than ever. This might well be caused by the fact that the kind of knowledge we need, does no longer fit very well with our common understanding of what knowledge entails. In this contribution, I hope to trace this gap back to its source and offer some suggestions as to how to bridge the gap between the traditional view on knowledge and the knowledge that is actually needed to face current and future challenges.

In the quest for knowledge, specifically scientific knowledge, the concept of rationality plays a crucial role. We are looking for rational, scientific solutions to the complex issues our societies are faced with. Talking about 'the' concept of rationality is rather misleading though, since the meaning of the notion has changed considerably in the course of time. About two centuries ago, rationality had become strictly limited to the cognitive aspect, i.e. the mental activities involving the processes of learning, observing, thinking, interpreting, and problem-solving. Only those explanations and solutions that resulted from observations and experiences acquired by the controlled use of our intellectual capacity were considered to be rational. People's everyday logic got into the bad books, as did the investigation into particular cases that could contain interesting insights and experiences. While in former times these kinds of specific, personal and context bound experiences were considered a fruitful source of knowledge, at the end of the 17th century the emphasis was put on timeless and context transcending knowledge.

Consequently, new demands were formulated with regard to the notion of rationality. Following these demands, philosophy was appointed the task of searching for abstract, general ideas and principles which would make it possible to relate specific cases to a context transcending level and produce general knowledge. 'Rationality' became a label reserved for knowledge obtained via our senses and intellectual capacity only, i.e. to what we currently call 'cognition' (Toulmin, 1990: 30-36 & 41; see also Verschoor in Chapter 9 of this book). Observation and experiences acquired by this faculty of our mind was considered to generate absolute, certain knowledge and was therefore put in central position within science. The development of scientific rationality was not just seen to form the basis for the progress of science, but for the progress of mankind as a whole.

Within this process of knowledge development, purity, exactness and strictness were the key words. Guiding principles were the separation of body and mind, the idea that the world is determined by general binding laws and the idea that emotions are a disruption of reason and therefore must be pushed aside. Together, these basic assumptions describe a system of ideas that can be viewed as the framework of modernity.

One narrow-angle concept of rationality, three modes of knowledge

Though there certainly are some concessions to be made, the 'modern' conception of rationality can fulfil a good deal of its purpose for questions related to mode 1 knowledge. Mode 1 refers to cognitive-instrumental knowledge, which more or less coincides with traditional, monodisciplinary forms of knowledge (i.e. knowledge concerning one singular scientific subject field) aimed to bring about objective, value-free and universally valid knowledge. Its central focus is finding explanations for how certain processes within a well-defined subject area work (e.g. biology, chemistry, psychology, see Figure 1). Within mode 1 science the academic community decides which questions are important and what answers are adequate. The knowledge and methods that are produced and applied are specifically suited to deal with that kind of question (Gibbons *et al.*, 1994; Nowotny *et al.*, 2001).

For other kinds of questions, these specific types of knowledge and methods have only limited relevance and value. But it is exactly those other kinds of questions that are currently and urgently pushing themselves to the fore. Many of the challenges we are currently facing imply not only cognitive questions, but also relate to matters of policy and to political, moral and existential insecurities. This book gives an impression of the kind of issues at stake here: the research project of Keune *et al.* (Chapter 2) focuses on the relationship between environmental pollution and human health and asks what policies should be designed to deal with this relation; Verschoor (Chapter 8) reports about the debate on genetically manipulated maize in Mexico and how different parties fight for their rights on this subject; Beukema and Valkenburg (Chapter 4) try to find ways, together with professionals and management, to bring about a demand-driven care for the elderly. And these are only a few examples.

<table>
<tr><th>Mode 1 knowledge

Cognitive-instrumental knowledge i.e. knowledge generated to find theoretical explanations and technical solutions for scientific problems, usually monodisciplinary in focus.</th><th>Mode 2 knowledge

Knowledge aimed to contribute to –preferably scientifically founded– solutions for practical problems and challenges which are characterized by their inter- or transdisciplinary nature.</th><th>Mode 3 knowledge

Knowledge concerning personal learning processes in which also moral and existential issues are addressed.</th></tr>
<tr><td>Narrow-angle rationality
Only the cognitive input, i.e. knowledge obtained via controlled observations and purely logical reasoning is considered to be relevant to find rational explanations and solutions for the problems as defined by the scientific community.</td><td></td><td></td></tr>
<tr><td colspan="3">Wide-angle rationality
Besides cognitive aspects like meanings and matters of facts, also norms and values, self images and power relations are considered to be relevant for finding rational explanations and solutions for the problems as defined by science and society.</td></tr>
</table>

Figure 1. Different types of knowledge and rationality.

If we know the causes of certain problems, what policy instruments can we develop to solve the problem? Undoubtedly, limited resources force us to make choices about which instruments to use or even which problems to solve. Our choices also reflect the stances we take with respect to the world we live in and the relationship we develop with nature and each other. Critically reflecting on our attitude and making sound decisions about these matters demand just as much rationality, be it in a broader sense. Due to the narrow conception of rationality, these issues have remained outside our view for a long time.

Certainly in western society, the dominance of cognitive-instrumental rationality hinders the other forms of rationality from entering the picture (Habermas, 1981: 102, 485). The tubular perspective that is responsible for the narrow rationality concept that has reigned since the end of the 17th century obstructs a wide-angle view of science. Every problem passes through a cognitive filter before it is 'scientifically' handled, which is still determining to a large extent the way and direction in which those solutions are sought and found (Kunneman, 1986).

To get a clear picture of how science functions from a social perspective, we must study the effects of the restrictive character of this tunnel view and its accompanying rationality with regard to both the material reproduction (i.e. the production of goods and food, etc.) and the symbolical reproduction of society (like values, culture, etc.). This analysis shows us that the rationalisation process set in motion two centuries ago actually has a double character.

On the one hand it forms a means to efficiently handle issues of a cognitive-instrumental nature and enables some of us to lead a comfortable life. On the other hand, the limited 'rational' outlook on life and the events, chances and challenges it has in store for us has serious drawbacks in the form of unwanted, undesirable side-effects. The negative effects of human interventions on our planet's nature is just one, but hopefully convincing, argument here. It doesn't seem very 'rational' to ruin the one and only habitat that we have. Moreover, what we see is that the fortunes have not only merely befallen the 'happy few' up until now, but also that it is mainly the unfortunate ones who have to pay the price of the negative side-effects the one-sided rationalisation process has created.

In short, the rationalisation process has brought far-reaching changes that can lead to domination (e.g. of human over nature or of humans over other humans), but can form a basis for liberation and emancipation too (e.g. releasing us from daily chores regarding our material reproduction or from oppressive social relations). The analysis of the double character of the rationalisation process enables us to avoid a biased view that would lead us to emphasising only the dangers of dominance or the alienation it incorporates. Just as it prevents us from exclusively focusing on the promises of liberation and the integration it holds. To correct the one-sided rationalisation, the modernisation process has up until now been characterised by the idea that we need to broaden our scope and open up our eyes to other aspects that should be taken into consideration if we want to live a truly rational life.

To begin with, besides mode 1 knowledge, there are several other knowledge modes that can be distinguished. At the turning point of the century, Gibbons *et al.* (1994) and Nowotny *et al.* (2001) introduced mode 2 knowledge. Mode 2 refers to practical knowledge, which is aimed at finding scientifically legitimised solutions for practical questions. These questions usually have an economic background or are related to policy matters. Such practice-oriented mode 2 knowledge is characteristically generated in multi- and interdisciplinary cooperation. Moreover, other stakeholders, besides the scientific community, play an important role in the assessment of the validity and relevance of the proposed solutions. Stakeholders are governmental organisations and private partners, but also NGO's and citizens. Mode 2 is a response to the needs of both science and society (Gibbons *et al.*, 1994: 11). It builds upon the results of the production of mode 1 knowledge, while in turn further development of mode 1 is also dependent on mode 2. But it is mostly judged on its practical applicability by a broad range of stakeholders, with all the tensions and conflicts that come along with it.

Beside mode 2 knowledge, Kunneman (2005: 116 ff.) distinguishes yet a third kind of knowledge. Mode 3 knowledge refers to intuitive knowledge; it is aimed at the elaboration of existential and moral insights in view of practical questions and dilemma's. Mode 3 learning processes do not usually follow a linear line. Instead, they have their own logic and are dependent on a certain friction between different points of view. Besides a variety of perspectives, mode 3 knowledge also calls for the input of inspiring narratives, visions and metaphors with regard to moral dilemma's and the big questions in life. As such, it forms the source that we can draw upon when dealing with the practical issues we are faced with, as well as with the inevitable tensions related to them.

The knowledge paradox: lots of knowledge, lack of innovation and change

There is still hardly any recognition of or place for mode 2 and 3 knowledge within universities. Though technical universities may acknowledge their role in our mode 2 knowledge society, it is usually left to the upper vocational education institutions and the publicly funded non-academic research institutes to bridge the gap between science and society (Canton *et al.*, 2005: 17). In general, universities cannot be said to have actively taken up the production of mode 2 knowledge – let alone mode 3 knowledge. They seem reluctant to give up their ideal of an independent science, committed to generating objective, value-free and context transcending knowledge.

The situation hasn't always been like this. In the seventies, the political dimension and social relevance of research programmes were strongly emphasised at universities, at least in the North. Nowadays, the pendulum seems to have swung to the other end of the scale. True, subsidy programmes place some demands on the involvement of people in the field of study and with regard to dispersion of research findings. But in the end, research production is exclusively defined in terms of academic output. The decisive assessment criterion of scientific quality is publication of the research results in highly specialised international journals that will in genaral be read by a few dozens of colleagues at the most. In view of the kind of demands related to this sort of publications – like those with regard to conditions about the level of control and the replicability of the research and the universal validity of the research claims – not many researchers commit themselves to projects related to mode 2 and 3 knowledge. Or at least there don't seem to be many scientists that have done it and managed to save their academic career.

Consequently, the gap between university and society is still yawning. In Europe this has led to a rather inconvenient and embarrassing 'knowledge paradox'. Universities and public research institutions generate an impressive innovative potential, but the countries do not know how to make optimal use of this public knowledge pool. Policymakers worry about the limited science-to-industry spill-over, including the transmission to the manufacturing and services sectors. Though the emphasis is usually put on the technological potential, the potential for innovations in the social and cultural domain should not be underestimated. The underutilisation by firms of the public knowledge pool reduces the possibilities for economic growth and the potential for innovation, since advances in productivity are driven by research and development (R&D) and by the use of results from public research. Therefore, many countries are reflecting on policies that enhance the transfer of knowledge from universities to companies (SER, 2003: 37; CPB, 2002, see Canton *et al.*, 2005).

The 'ill-designed reward structure for scientists in the public research sector' is identified as the most important barrier to an effective knowledge transfer (Canton *et al.*, 2005: 11). The 'publish or perish' culture, i.e. the pressure to disclose the knowledge produced in international publications is really the only incentive scientists experience. This scientific 'norm of disclosure' does not always marry well with the norm of secrecy common to the business world (though there are also signs that science and business mutually benefit from scientifically accredited publications). Another barrier to the transfer of knowledge is a lack of absorption capacity of private companies. Absorption capacity is a function of a firm's investments in R&D and its relations with the scientific world. A company investing in R&D and in networks with the academic community can benefit more from the public knowledge pool. But empirical research (CPB, 2005, relying on Eurostat CIS-3 2004) shows that very few companies mention universities as a significant source of information, not just in the Netherlands but in EU countries in general.

Additional barriers in the exchange of knowledge are diverging research agendas in the public and private domain. Of course, considering the public tasks of universities and the anticipated knowledge spill-over, some differences between the research topics within public research institutes and the R&D centres of private companies are legitimate (*ibid.*). But if we want to enhance the transfer of knowledge, universities should be more tuned in on the demands of market and society when designing their research agendas.[62] The existing structure of research programmes forms an impediment to enhancing the creation of complex and practice-oriented research questions (i.e. mode 2 objects) around which knowledge is developed. For one thing, the common distinction between a programme commission exclusively manned by scientists and a steering group formed by policy makers does not enhance the interaction that is needed to establish more converging agendas. Moreover, calls for tenders commonly used to assess the eligibility of research programmes cannot accommodate integral approaches. The criteria for judging research proposals, with their strong focus on scientific excellence in a certain discipline, tend to exclude multi- or interdisciplinary research programmes (cf. RMNO, 2005: 12-13, 21-22). Fortunately, there have been initiatives over the last couple of years to stimulate more multi-, inter- or transdisciplinary research (e.g. in the Netherlands the WOTRO-programmes of the Dutch Organisation for Scientific Research (NWO) and on a European level Edulink). But it is still quite difficult to find and create opportunities for research that transcend the traditional, disciplinary organised scientific structures within academia.

[62] Tuning in to demands from market *and* society, for it's not just science-to-industry spill-over that needs to be taken into consideration of course. Private companies can have a positive effect on the wealth and well-being of a country's population with their innovative activities, but the interests of society as a whole do not necessarily run parallel to those of singular commercial companions.

Parallels between the North and the South

While profound changes are taking place in the European university world, the universities themselves have hardly changed. As the Commission of the European Communities remarks: 'After remaining a comparatively isolated universe for a very long period, both in relation to society and to the rest of the world (...), European universities have gone through the second half of the 20th century without really calling into question the role or the nature of what they should be contributing to society' (CEC, 2003: 22). And what is happening in Europe is not unique. As Kibwika (2006: 20, 42-49) observes, universities in developing countries also remain 'ivory towers' for academics far detached from the development process. Higher education as it exists today in Africa is to a considerable degree a reflection of higher education in the North, i.e. Europe and the USA (the land grant universities). It is based on the same models, it is faced with comparable changing conditions at the beginning of the 21st century and it is also struggling to meet the new demands. If there is one reason to be careful when drawing a parallel, it is only that the challenges are often much greater in Africa, Kibwika rightfully notes.

African universities and the education and training they offer have been de-coupled from potential resources for national development and wider African social values right from the start, at least in East Africa. Starting as a technical college in 1922, with the approval of the then Secretary of State for the Colonies, Winston Churchill, Makarere became a university college supervised by the federal University in London in 1937. The underlying motive for this transition from an emphasis on the 'hand' to an emphasis on the 'mind' was the need for 'native' white collar workers for the colonial administration (Ssebuwufu, 2005 in Kibwika, 2006: 42).

As one of the major instruments and vehicles of cultural westernisation on the continent, African universities were a driving force in the 'colonisation of the mind' (Tiberondwa, 1998 in Kibwika, 2006: 43). Uncompromisingly foreign in an African context, the impact of the original colonial university was more culturally alienating than it need have been (Mazrui, 2003 in Kibwika, 2006: 43). Rather than equipping Africans with the capacity to develop themselves, education on the African university enforced patronage to the colonial masters. When the universities in Africa became independent, they still continued to offer a type of education based largely on European models. Even in the 1970s and 1980s, when the direct colonial influence on the universities was finally disrupted, they did not manage to innovate on the institutional level. The education and research training was still modelled around principles of hierarchy and bureaucracy that had not changed since the colonial times (Kibwika, 2006: 43-44).

So it might be clear that in Africa too, the rationalisation process reveals its double face. Perhaps it is even difficult to see any positive side of it, considering all the downsides related to the one-sided rationalisation that was introduced by colonialism and that is still holding the universities in its grip. By-and-large, the post-colonial African academy was expanded without much or any attention to the quality improvement needed to suit the new needs and challenges (Byamungu, 2002 in Kibwika, 2006: 43). Therefore, Kibwika argues, it is necessary that African universities critically review the needs, challenges and opportunities that lie before them. 'It is no longer in debate whether the African university should change; it is rather a question of 'how'.' (Kibwika, 2006: 48).

To start with, the universities should explicitly articulate their societal responsibilities in vision and mission statements. But this is not enough; it has to be accompanied by processes and mechanisms for bringing universities nearer to society and to the challenges of poverty alleviation. Up till now, the changes in the curricula we would expect to be brought about to enhance the implementation of the vision and mission statements, are still non-existent. Put another way: the new social contract has to be operationalised, i.e. to be translated into concrete action plans and must not remain just a dream on paper. Though the background might differ, European countries are facing the same task (cf. AWT, 2003).

10.3 Research as interactive learning processes

Intervention research, action research or any other kind of utilisation-focused research all have the aim to change, innovate and improve the existing situation. Within this specific form of research with its intervening, process-oriented character, the design, management and evaluation of the intervention or action strategy form an integral part of the knowledge production. Within this type of research, mode 1 knowledge as well as mode 2 and 3 knowledge are developed and employed. As such, it forms a means to help bridge the current gap between what traditional science picks up and the actual problems society is facing.

A starting point within intervention research or action research is that the production of knowledge implies an active knowledge creation process. Which methods are preferred and actually being used within this process is dependent upon the principles guiding it. Since no single explanatory model can declare itself valid and no single theory of science can predict what value it will have for mankind, there is neither an absolute set of principles nor just one scientific method. Though logical empiricism – also known as the empirical analytical approach – used to be viewed as the 'standard model' in the orthodox consensus, it is questionable whether this model is suitable for generating other modes of knowledge beside the traditional mode 1 knowledge. Moreover, the claim that the scientific knowledge generated by empirical analytical research would be objective, value free and universally valid has been severely criticised by philosophers of science (cf. Tromp, 2004, 2008), as its favoured methods – observation and logic – are not free from flaws. The process of making observations and of forming statements based upon these observations does not just simply refer to reality (as is already indicated by the active verbs 'making' and 'forming' related to these acts), but also implies certain theoretical presumptions that lead the way. These presumptions form our perspectives, direct our observations and colour our interpretations. So 'vision' is never a neutral act; it is prompted by basic theoretical, or if you wish, philosophical ideas.

According to Gibbons *et al.* (1994: 105) 'the scientific community and also its analysts have traditionally emphasised the relative autonomy and the functioning of science as a distinct subsystem of society. Largely accepted until the 1960s, this has been contested ever since. It does not account well for the practice of science even in mode 1, and it certainly does not obtain for mode 2.' In short, the traditional scientific research model, with its exclusive focus on the cognitive aspect and mode 1 knowledge, falls short in terms of offering an adequate directive and overarching framework. To get a clear view on a more suitable model, another perspective and another paradigm of knowledge production are needed (cf. *ibid.*: 101). Below I will try to sketch how we could try and adjust the perspective to one that is more suited to accommodate a wide-angle approach to science.

From a linear to a cyclical model of knowledge creation

In the traditional linear model, knowledge production is considered to take place in more or less well-defined stages - from knowledge development to the application of knowledge. This linear perspective, which lies at the root of the failing technology transfer model, can no longer be held. In this model, knowledge production is presented as a relay race in which the baton (knowledge) is passed on by each subsequent runner in the game. Thus, it would move from fundamental science to applied science to product development, to finally end up in the phase of commercialisation. In fact, the creation of knowledge should rather be viewed as a cyclical process of change. And innovation or social change, consequently, as a spiralling process in which knowledge generation, application and management intertwine in a very subtle, complex way (cf. Figure 1 in the contribution of Ramaru *et al.*, Chapter 3 in this volume. See also Beukema and Valkenburg, Chapter 4 in this volume).

It is increasingly acknowledged that knowledge is context dependent, is highly dependent of creativity and 'tacit knowledge' (Gibbons *et al.*, 1994: 17), i.e. knowledge that cannot be laid down in reproducible forms like books or computer programmes. Beside skills, tacit knowledge involves elusive forms of knowledge in people's heads like knowledge from experience, expertise and intuition. It is recognised that innovation relies upon the work of people and is a result of the (apparently often highly arbitrary) interactions between the various stakeholders. Since it is crucial that the knowledge circulation process develops fluently, barriers between the various links within the innovation chain must be removed. These links should be adjusted to each other early on in the process. And each separate link should be working well, as 'the most marvellous inventions will not lead to innovation when there are not enough competent persons to bring these inventions a step further by up scaling, marketing and production' (SER, 2003: 20, 29 & 59; HOOP, 2004: 11 & 20).

Certainly, this new perception offers better chances to incorporate the different types of knowledge and the wide-angle rationality concept that are needed to develop promising innovative perspectives. But still, there is a tendency to perceive the process of knowledge generation as a kind of production line that needs to be organised as effectively as possible. Innovation is regarded as a renewal - of commodities, services, processes or organisation forms - that calls for new knowledge, either generated by doing research or acquired by combining various forms of existing knowledge (SER, 2003: 19 & 58). Valorisation is defined as the creation of societal added value on the basis of academic knowledge (OC&W, 2005: 2).[63]

What is missing in these definitions becomes clear when we compare this perception with the definition offered by Kibwika (2006: 166): innovation is an adaptation and translation of learning into options for solving real-life problems in a complex and dynamic environment. Change is subsequently viewed as the desirable outcome of learning and innovation to improvement of life (or better development). Put differently, we must stress the dynamic dimension of knowledge creation as a result of the interaction between scientific researchers, community members and other stakeholders as part of a dynamic and multi-layered learning process. This is also what Gibbons *et al.* (1994: 87-88) propose in their plea for a new paradigm of knowledge production: 'The older view of a linear process connecting discoveries and inventions to the production process is displaced by a more interactive one. While in the linear view, the university was distanced from the commercial process, and could still preserve its academic values, in technology interchange it must become involved at both individual and institutional levels and adapt to new rules.'

First and foremost we must realise that valorisation, i.e. making knowledge of value for society, implies more than just the transmission of knowledge. It is not simply a matter of passing the knowledge baton from one runner to the next; the players really have to form a team. Scientific knowledge can only then become socially robust knowledge when the knowledge production within society is considered to be transparent and participative. This relies on a reciprocity in which the general public understands how science works, and the other way around: science also understands how (re)actions of the various groups stakeholders and the general public come into being (cf. Tromp, 2004, 2007 on the importance of reciprocity within research that is aimed to bring about change; see also Beukema, Chapter 11 in this volume). This reciprocal understanding must be led by a vision that needs to be explained in a language that is understood by a larger audience. This vision is process oriented; it does not contain a blue print (Nowotny *et al.*, 2001: 248-249). Such a vision is based on

[63] With this definition, the Ministry of Education, Culture and Science (OCW) applies a correction to the narrow definition that was formerly given in the Science Budget (2004), in which valorization was described as: "converting the results of scientific research into economic value" (OC&W, 2004: 11). It seems to be illustrative for the change in mindset that still has to be made.

theoretical and philosophical assumptions and is not just the result of objective and value-free observations, as will be further argued below.

A methodology as a visionary framework

A research model and a methodology are usually seen as one and the same thing. They are taken to be overarching notions for a 'tool kit' of research methods, techniques and instruments. This is a rather limited conception of the notion of 'methodology' though. In a broader conception, a methodology is also regarded as a framework for analysis that contains the rules about how (a specific form of) research is to be done. It forms a system of logical or philosophical principles against the background of which theory relevant data are structured. Subsequently, a coherent set of research methods is developed.

In this conception, a methodology is regarded as a theoretically led research framework, founded on several presumptions regarding the research object (things and phenomena in the physical and social reality). Within this framework, we are faced with questions like: what vision on reality guides our view when we do research, what do we see as the constituents of reality? What kind of perspective directs our view on human beings, or put differently: what power and forces do we attribute to the actors involved in the research? How can we gain understanding of a continuously changing reality? What kinds of knowledge are exactly needed to be able to do so? How can we find explanations for the success or failure of the solutions or strategies we come up with? What is the relevance of something that 'works' or doesn't work in one situation for another situation? How do we deal with the social responsibility for the effects of a scientific solution or intervention, that might generate unintended and undesirable side effects or that might not change anything at all?

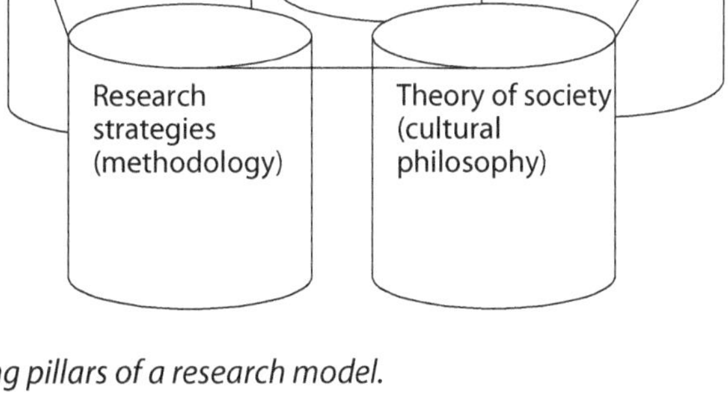

Figure 2. Underlying pillars of a research model.

Considered from the viewpoint of philosophy of science, these are questions that subsequently refer to: ontology (i.e. the view on reality), philosophical anthropology (i.e. the view on human beings), epistemology (i.e. the theory of knowledge), methodology (i.e. the view on what is the best way to do scientific research) and cultural philosophy (i.e. the view on society and the place of science within it, cf. Figure 2). It seems inevitable that we explicitly reflect and are open about the presuppositions we hold on these fundamental issues, since there is no way we can avoid relying on such presuppositions.

No scientific approach or research model can do without these pillars of knowledge creation (cf. Tromp, 2004 for an elaborate argumentation to back this claim). Therefore, I will take the broader conception of methodology as a starting point, which means that we have to deal both with theoretical and more practical aspects related to any methodology. We have to consider questions like: what kinds of research conceptions can we distinguish and how can we choose a research strategy that is suitable for the challenges we are faced with? What (re)sources can we rely on in our quest for suitable strategies and solutions? How can we make use of these resources? And what kind of competencies do we need to do that? Gradually shifting from the theoretical to the more practical level, I hope to be able to shed some light on both types of questions.

To develop a vision a wide-angled concept of rationality is needed

The recognition that every methodology is necessarily built on certain pillars and basic underlying principles clarifies why we really need a broader conception of rationality. This applies to all research models, no matter how a-theoretical or a-political they claim to be. It certainly applies to research and research models that are meant to make a difference and to develop future oriented visions.

In research aimed at inducing change, there are different actors and factors, situation-specific and situation transcending goals, interests and perspectives to be dealt with. Prevailing values and norms will have their impact on the decision to 'go for' a specific strategy or intervention or to refrain from it. Whether an action plan is actually going to be implemented is not solely the decision of scientific analysis, but in part depends on the dominating power relations in society. Moreover, the aims related to the planned interventions and the criteria used to evaluate them will be worked out differently by representatives of different parties (e.g. political parties). Which aims and criteria are given priority will also depend on the social and political constellation. Since their interests will not always run parallel, it is inevitable that tension will sometimes exist between the different stakeholders (cf. with the different 'regimes of justification' that Verschoor distinguishes).

In addition, the cultural factor evidently plays an important role as well. This applies both internally and externally, i.e. it can be said about our multicultural singular societies as well as for our global society. It's important to be aware of in- and excluding effects of interventions and action plans and of research approaches aimed to bring these about. It's also important to make room for differences.

All in all, it is clear that more aspects are implied than merely cognitive-instrumental aspects. The technological or empirical analytical research model starts from a rationality concept in which there's actually only room for cognitive-instrumental aspects. When we want to bring into view a research model that can also do justice to the meaning and importance of norms, values, morals, (political) sensitivities and interests, this calls for an approach in which a wide-angle instead of a narrow-angle concept of rationality is used. Elsewhere, (see Tromp, 2004, 2008), I have tried to show how such a wide-angle concept of rationality could be developed in the last quarter of the 20th century. For after the crumbling of the orthodox consensus about the traditional, narrow-angled concept of rationality, several turns have taken place within the philosophy of science that provide a fruitful input for a rationality concept that incorporates a broader perspective. Within such a rationality concept normative aspects, aspects of power, intuition and emotions are also taken into consideration besides meanings and matters of fact (cf. Figure 1).

A wide-angle rationality concept alone is not enough though. We also need a theory of knowledge that provides room for such a concept. The traditional paradigm in which knowledge is seen as the product of an objective, individualistic, a-historical procedure cannot accommodate such a concept. But a paradigm in which knowledge creation is viewed as an interactive process in which researchers put their knowledge constructions to the test, can. Within such a concept of science, matters of fact, normative issues and power relations can all be made into a topic of discussion to find out what is the rational thing to do. As such, they can all become subjects of investigation within the process of rational knowledge creation which takes on the form of an argumentative learning process.[64] Reflexivity plays a crucial role in this process, as will become clear below.

10.4 Building bridges between science and society

Transcending dualistic thinking by opposing oppositions

To start with, a reflexive attitude is necessary to distance ourselves from untenable conceptual dualisms that still often mark scientific theorising and research practice. Within dualisms, concepts are put opposite each other and presented as pole and antipole, as direct oppositions in a dichotomy. The central dichotomy of modernity, in which corporality and intellectuality are put opposite each other, has been the source of many more dichotomies. An example

[64] It must be emphasized here that the backing, the legitimation of validity claims concerning norms and values or power relations or self-images is of a different character than that of meanings and matters of fact. Therefore, the backing takes on another form. Moreover the authority and power of these different claims can vary. Within the scope of this chapter I will not be able to elaborate on this point though. The interested reader is referred to Tromp (2004), chapters 5 and 6.

is the idea that emotions are a disruption of reason – one of the other important pillars within the framework of modernity. But also the opposition of subject and object, facts and values, knowing and acting, theory and practice, objective structures and subjective actions, determinism and voluntarism, applied and fundamental research, the abstract and the concrete, the universal and the particular et cetera.

Of course, it is inevitable that we make choices in the focus of our research projects and the concepts and approaches that we choose, since our time and resources are limited. But we must realise that dualisms, with their exclusive 'either/or' character, imply theoretical distortions of the complex reality (Bourdieu, 1998). Focusing exclusively on one side or the other cannot give an accurate picture. And most of the time, it will not do just to 'add up' findings from both sides to get a proper view on things. Rather, it's like a dialectic process in which thesis and antithesis have to be brought into synthesis. Therefore, we should try to overcome thinking in terms of simplistic dualisms. For not only does it bear the risk of theoretical distortion; thinking in terms of oppositions doesn't help building bridges either.

One particular way of thinking that needs to be reconsidered is the model of systems thinking that is very influential in the physical as well as in the social domain, both within science and in the world of policy making. What is enticing in these 'systems' type of models, but what also makes them rather treacherous, is that they 'imply greater stability in the relationships between actors than is justified given what we already know of distributed knowledge production.[65] More helpful descriptions could be worked out by trying to develop models that incorporate the evolution of patterns of interconnections, the ability to establish, on a recurrent basis, new modes of exchange, the skills to adapt to the richness of research practice, and to create ever new channels of communication' (Gibbons *et al.*, 1994: 155 & 160). For example, we can only study the effects of soil fertility management on crops when we know what role the farmers in the field play within the system of rural small-scale farming as well as the persons they depend upon for trading, etc. (cf. the contribution of Ramaru *et al.*, in this volume). Likewise, we can only gain insight into what happens in the process of curriculum development within (a specific domain within) academia when we take into account the actions employed by both teachers and students, placed against the background of how the university system works (in the particular case discussed here the Wageningen University, cf. Almekinders *et al.*, in this volume).

Structuration theory (Giddens, 1976, 1979) can be a source of inspiration to help free ourselves from the dualisms that are still holding theory development and policy making in its grip and hinder a fruitful interchange between science and society. Within structuration

[65] Gibbons et al. (1994) refer here to the new kind of knowledge production, which does not take place in one particularly place – like the university – but happens in different locations, within networks and relationships of people. Hence they speak of distributed knowledge production (see also the next paragraph).

theory, the opposition between 'action' and 'structure' is overcome, since it is exactly the interrelationship between the two that is put in a central position. By taking structuration as our starting point, neither structure nor action will be the sole subject under study. The attention is both focused on the influence of more or less stable structures (that together can form a system) on the actions of the people in the specific research field and the other way around, i.e. on the influence of people's action on the system. Consequently, it is by definition impossible for research to remain limited to just one perspective. An additional benefit of this approach is that other dualisms can also be avoided, for instance the opposition of determinism and voluntarism. This way, we can prevent ourselves from tumbling into the pitfall of either 'blaming the system' (the deterministic bias) or 'blaming the victim' (the voluntaristic bias) when things do not work out the way they were planned.

Things are (even) more complicated when the object of study pertains to the domain of the natural sciences. For it will be even harder to find out in what way structures and actions are related and what kind of exchanges take place on the interface between physical causality and human influence. For instance, we can assume that system earth and its subsystems of the biosphere, hydrosphere, geosphere and atmosphere will function whether human life had developed or not. But since we, human beings, are here and rove about the earth, we seem to have quite an impact on the planet. To what extent human actions are the cause of changes in, for instance, our climate is very hard to determine. But focusing on one aspect (the way the various subsystems interrelate) at the cost of attention for the other (the influence of human actions on this complex of interrelations) will definitely not get us where we want to go.

The need for an interdisciplinary perspective and integrative approach

Just as dualisms are theoretical distortions of the complex reality, so are approaches in which one perspective is pitted against another. Investigating phenomena from only one perspective in general cannot do as much justice to their complexity as an interdisciplinary approach can. So, a second issue related to the need for reflexivity is the question of how justice can be done to a holistic approach to the object or co-subject of study, i.e. how to determine what disciplines and perspectives should be involved in the query.

Since current, actual issues cut through various domains and disciplines, research cannot usually be limited to only one field or discipline. As the Commission of the European Communities (2003: 17) rightfully acknowledges: 'advanced research increasingly falls outside the confines of single disciplines, partly because problems may be more complex, more because our perception of them has advanced, and we are more aware of the different specialisations required to examine different facets of the same problem.' In view of the need to develop interdisciplinary capability, the Commission formulates as a condition for excellence 'that universities be enabled, and encouraged, to develop more work falling between the disciplines. Organising work on an interdisciplinary basis requires that universities have flexibility in their organisation, so that individuals from different departments can share their knowledge and work together (...). It also requires flexibility in the way careers are evaluated and rewarded, so that interdisciplinary work is not penalised for being outside normal departmental frames. Finally, it requires that departments themselves should accept 'cross-border' work as contributing to faculty-wide objectives'.

Instead of monodisciplinary research, we need a poly-angulated approach. Even a multidisciplinary approach may be insufficient, because the various perspectives are merely set beside each other. In an interdisciplinary approach on the other hand, perspectives are combined and integrated by viewing them in their mutual coherence (or incoherence, if there is a problem!). Within multidisciplinary research, synthesis of the different disciplinary contributions takes place post hoc, when the individual outcomes of disciplinary research are combined within a problem solution. Interdisciplinary research requires a more integrated approach throughout the research process. It requires a joint research approach by researchers from different disciplines. Different disciplinary conceptual frameworks might be used but problem definitions, methodology and interpretation of results are dealt with in joint consultation (cf. RMNO, 2005: 8).

Transdisciplinarity still goes a step further; it can be viewed as a specific form of interdisciplinarity in which boundaries between and beyond disciplines are transcended and knowledge and perspectives from different backgrounds are integrated. Transdisciplinarity is a characteristic feature of mode 2 knowledge (Gibbons *et al.*, 1994: 17). The goal is to develop knowledge that is relevant for the solution of complex societal problems. 'As a consequence, this dispersed and transient way of knowledge production leads to results which are (...) highly contextualised' (*ibid.*). Depending on what is needed, researchers with varying backgrounds are employed to help generate the necessary knowledge. The sources relied on are not restricted to the scientific domain. Non-formal knowledge, like knowledge based on practical experience and knowledge not related to any particular scientific discipline, is also being used. Moreover, within a transdisciplinary approach the research question is not solely determined by scientists, but is formulated in a mutual interactive and iterative consultancy between the parties involved (RMNO, 2005: 8-9). Since the knowledge that is created is embodied in people and the ways they are interacting, transdisciplinary knowledge is not confined to the context it is generated in, but becomes socially distributed as the 'knowers' move on to different contexts of application or use (Gibbons *et al.*, 1994: 17).

Such an inter- or transdisciplinary approach matches well with the alternative rationality concept, methodology and paradigm that are proposed here. But as we have seen, the existing structure of research programmes and their criteria for (e)valuation as yet hinder rather than enhance the development of such integral approaches.

Combining different kinds of knowledge

In research aimed to bridge the gap between science and society and to bring about innovation and change, the relationship between scientific theory and daily practice must be taken into account. This means that one has to be explicit about the goals of (a specific) research, as well as about the influence of the researcher on the object of research. Within this context, reflexivity presupposes explicitness about the background of the research and the presuppositions and values that are held regarding the research object. This pertains to questions like: What presuppositions and assumptions are held with regard to the research object (as can be partly deduced from the interpretative schemes that are used)? What knowledge can be relied on and what knowledge is still missing about the issue under study? What arguments can be introduced for using certain specific research methods and methods for change? Why are some choices made in the learning cycle of change in general and the research in particular rather than others? These demands to be explicit about where one stands applies to the researcher, but to the researched as well! Put differently: reflexivity is needed from both the researcher and the researched (Coenen, 1998).

A methodology that takes into account both mode 1, mode 2 and mode 3 knowledge implies a principal equivalence between the knowledge and competencies of researchers and those of the researched. This equivalence should also find a translation in the role that both play within the process of knowledge gathering. The basic principle is that as the researcher can be an expert on a specific scientific field of interest, so the researched can be considered to be experts when it comes to their own situation. Therefore, we have to abandon the idea that the researched are no more than naive sources of information about that part of reality we want to investigate. Their practical knowledge, their knowledge from experiential learning and sometimes even theoretical knowledge of the particular context need to be incorporated in the knowledge creation process. And their assessments and validations of the created knowledge have to be taken seriously. In the end, knowledge can only considered to be adequate when it is recognised as such by both researchers and researched (*ibid.*).

These implications can be met by setting up research as a mutual learning process in which the researched become co-subjects in the research process. A learning process in which the knowledge of both subject (researcher) and object (the researched) are regarded as indispensable, complementary in the research process and as decisive for the quality of the results (*ibid.*). The researchers put their validity claims to the test in critical discussion, not just with their colleague scientists, but also with those involved in the actual research project.[66] By doing so, the various forms of knowledge that play a role in innovation and

[66] Within the traditional approach the role of the forum remains restricted to fellow scientists and really cannot be accounted for, since the underlying theory of knowledge is founded on the principle of a disinterested individual researcher that makes objective, theory- and value free observations and not of a scientific community that negotiates about knowledge in an interactive critical debate.

change processes and the expertise of the relevant actors and stakeholders with regard to these different kinds of knowledge are acknowledged.

Then again, in research focused on innovation and change the emphasis is on the action that needs to be undertaken to produce a successful improvement strategy in the situation at hand. The empirical dimension is surely not lost from view. Indeed, experiments and the concrete, practical experiences of the actors and the context in which these (old and new) experiences take place form an indispensable dimension within the theoretical explanatory model.

Both universal expert knowledge and local contextual knowledge are needed to realise more sustainable solutions to issues that potentially have strong political, moral and ideological foundations. Citizens do not have to be active and empowered in every issue but can move from more passive detachment to active involvement at different times and vice versa (Wals and Heymann, 2004: 19; Tromp, 2004: 260-262; Coenen, 1998). Expert knowledge and skills come in to support a change process in motion, but they themselves cannot induce change in a social system. For that, deroutinisation of existing action patterns needs to take place and the current practices gradually need to be replaced by new practices. Expert knowledge is an essential but scarce public resource – hence the need to find the most effective way to utilise it (Kibwika, 2006: 170).

Scientists can fulfil an important task in enhancing the transmission, translation and valorisation of knowledge by building relationships on the intersection of lay- and expert discourse, including the scientific discourse (i.e. the discussion taking place within science, or of science with the general public, all over the world). They can bring together the various forms of expertise and practical knowledge of lay actors and use them as input for expert discourses (cf. Nowotny *et al.*, 2001: 247-248). Their expertise reaches beyond purely knowledge production; they can also play a key role in connecting knowledge to decision making and action (*ibid.*: 216). This concerns not only mode 1 matters but also issues that relate to mode 2 and 3 knowledge, like searching for solutions for practical problems and finding the right attitude with respect to existential questions related to life and nature.

To enhance participation and reciprocity in the knowledge creation process ample attention must be paid to deconstructing the traditional role expectations regarding the researcher. Moreover, the (new) role the partners will play in the research process will have to be clarified and accepted. A precondition to realising such a new situation is to invest in the building of trust between researcher and researched. Trust is the medium on the basis of which individuals are to interact with an expert system like science. It is a precondition for individuals to get into contact with the science system – often a 'dive into the unknown' - with hopeful expectations. The relationship between individual actions and expert systems is not characteristically defined by trust though; a sceptical attitude or rejection is equally possible.

In the end, it becomes clear that connection is the key within research processes that are particularly aimed at innovation, valorisation or social change. For an innovation or change process to be successful the overall story of the enterprise, i.e. the shared mission and the overarching vision, need to be connected to each individual element, i.e. the underlying projects, small-scale experiments or pilots. Moreover, the vision needs to be connected to action (cf. Beukema and Valkenburg, in this volume). The fusion of mode 1, 2 and 3 knowledge that forms the foundation for innovative change processes requires integration of different skills in a framework of action (cf. Gibbons *et al.*, 1994: 4; Kunneman, 2005).

So we need to develop research programmes with an improved performance with regard to the transmission, adaptation, translation and valorisation of knowledge from academia to society. Research that focuses both on theoretical understanding and normative reflection and offers scientific backing for actual issues in order to be able to develop future minded visions, scenarios and action strategies. Above all, we need scientists and researchers with the competencies to close the ever present gap between science and society. For that, we need researchers with competencies that will enable them to cope with uncertainty, poorly defined situations and conflicting or at least diverging norms, values, interests and reality constructions (Wals and Jickling, 2002: 124).

In this book, we try to explore the present situation and to bring to light ways of tackling current challenges and dilemmas. Its aim is to teach and train a new generation of scientific researchers that are fit to face the future. While the situation in the North obviously differs considerably from that in the South, all of us are confronted with comparable global issues and developments that force us to change things. Though maybe less pressing in the former context than in the latter, the challenge is inherently the same: how can we educate and train researchers to help reach the aim of improving on the societal benefits that can be gained from science? So let's join forces to help reach that goal and with that, hopefully, create a slightly better world.

References

AWT, 2003. Towards a new societal contract. Synergy between public knowledge institutions and the Dutch knowledge society. The Hague: Council for Science and Technology Policies. Advice No 50, 16 January 2003 (in Dutch).

Bourdieu, P., 1998. Practical reason: On the theory of action. Polity Press Cambridge.

Canton, E., D. Lanser, J. Noailly, M. Rensman and J. van de Ven, 2005. Crossing borders: when science meets industry. Netherlands Bureau for Economic Policy Analysis (CPB), document 98, 5 October 2005.

Centraal Plan Bureau, 2002. The pillars underneath the knowledge economy. Options for institutional renewal. SDU, Den Haag (in Dutch).

Coenen, H., 1998. On the foundations of a relationship based on equality between researcher and the researched party in exemplarian action research. In: B. Boog, H. Coenen, L. Keune and R. Lammerts (eds.), The complexity of relationships in action research. Tilburg University Press, Tilburg, pp. 17-36.

Commission of the European Communities (CEC), 2003. The role of the universities in the Europe of knowledge, communications from the commission COM, 58 final, Brussels, 05.02.2003. Available at: http://europa.eu/eur-lex/en/com/cnc/2003/com2003_0058en01.pdf. Accessed 8 February 2007. Available at: http://ec.europa.eu/education/doc/other/orient/orie_en.html. Accessed 8 February 2007.

Gibbons, M., C. Limoges, H. Nowotny, S. Schwarzman, P. Scott and M. Trow, 1994. The new production of knowledge: The dynamics of science and research in contemporary societies. Sage, London.

Giddens, A., 1976. New rules of sociological method. Hutchinson, London.

Giddens, A., 1979. Central problems in social theory. Action, structure and contradiction in social analysis. MacMillan Press Ltd, London, Basingstoke.

HOOP, 2004. Higher education and research plan. Ministry of Education, Culture and Science, The Hague.

Habermas, J., 1981. Theorie des kommunikativen handelns I & II. Suhrkamp Verlag, Frankfurt am Main. In English: The theory of communivative action. Beacon Press, Boston 1984-1987 (Translation by McCarthy, T.).

Kibwika, P., 2006. Learning to make change. Developing innovation competence for recreating the African university of the 21st century. Wageningen Academic Publishers, Wageningen.

Kunneman, H., 1986. De waarheidstrechter - een communicatietheoretisch perspektief op wetenschap en samenleving, Boom, Meppel.

Kunneman, H., 2005. Voorbij het dikke-ik. Bouwstenen voor een kritisch humanisme. SWP, Amsterdam.

Ministerie van Onderwijs, Cultuur en Wetenschap (OC&W), 2004. Science Budget 2004. Focus on excellence and more value. Ministery of Education, Culture and Science, The Hague.

Ministerie van Onderwijs, Cultuur en Wetenschap (OC&W), 2005. Valorisation of research as task of the universities. Letter of the Ministery of Education, Culture & Science to the Board of the Dutch universities, The Hague.

Nowotny, H., P. Scott and M. Gibbons, 2001. Re-thinking science. Knowledge and the public in an age of uncertainty. Polity Press, Cambridge.

Raad voor Ruimtelijk- Milieu- en Natuuronderzoek (RMNO), 2005. Interdisciplinarity and policy relevance in research programmes. Taking stance, The Hague: Council for Spatial, Environmental and Nature Research (RMNO), juni 2005 (in Dutch).

Social Economic Council (SER), 2003. Interaction for innovation, towards a coherent knowledge policy. SER, The Hague (in Dutch).

Tromp, C., 2004. Wide-angle science. A critical-reflexive research model based on a broad rationality concept. Jan van Arkel, Utrecht (in Dutch).

Tromp, C., 2008. Criteria for a reflexive approach to knowledge production, in: Boog, B., Preece, J., Slagter, M. and J. Zeelen (eds.), Towards quality improvement of action research: Developing ethics and standards, Sense Publishers, Rotterdam, Chapter 2, pp. 7-28.
Toulmin, S., 1990. Cosmopolis. The hidden agenda of modernity. The Free Press, New York.
Wals, A. and F. Heymann, 2004. Learning on the edge: Exploring the change potential of conflict in social learning for sustainable living. In: A. Wenden (ed.), Educating for a culture of social and ecological peace. State of New York Press, New York, pp. 123-145.
Wals, A. and B. Jickling, 2002. 'Sustainability' in higher education: from doublethink and newspeak to critical thinking and meaningful learning. Higher Education Policy 15: 121-131.

11. Unity in diversity

Many forms of action research

Leni Beukema

11.1 Introduction

This chapter focuses on choices in design and implementing research that every researcher faces, when she/he wants the research to contribute to social change. I focus specifically on action research (AR) as an approach in social science, where research, learning and action go hand in hand. In action research experience on possibilities and dilemmas in research for social change is built up for decennia. From these experiences other researchers can learn when making their own choices and gain ideas on possibilities and dilemmas that come with them. Action researchers have sought a multiplicity of answers for the questions they faced, also dependent on their own view on knowledge development and on the context of their research. This variety and the backgrounds of it will be dealt with in this chapter.

We have mentioned the general themes of the choices on research in action in Chapter 1 of this volume, when talking about the central characteristics of research for social change. These themes are further elaborated on in this chapter: how did action researchers deal with these choices and what reasons do they put forward for their choices? Firstly this concerns the component of change in the research, both the goal of the research (emancipation, empowerment, democracy) as well as the extent to which the change process itself is part of the research. Secondly I focus on the theme of the relationship between researcher and researched party/co-researchers, i.e. the role of the researched party in the research process, the nature of the relationship between researchers and researched party and the chosen methods that are characteristic of certain relations. Finally I will elaborate somewhat on a much debated point in AR: the question of reliability and validity (or, in terms of AR researchers, trustworthiness, quality and validity). This debate is between AR researchers and traditional researchers, but is also a central point of discussion within the community of action researchers themselves.

Action research is a much differentiated field, reinforced by the contextuality of research practices. For instance AR can be found in community development, agriculture, environmental studies, education, organisational development, social movements, etc. Each of these fields puts different demands and opens up different opportunities for the research. But also within this different field we see different forms of AR, dependent on the choice of researcher and researched party and on the practical situation in which the research is carried out.

Boog (2003) in his overview of current approaches within the AR paradigm distinguishes four main streams, which share some characteristics of AR (like research as a cyclical process, emancipation and democracy as goals of the research, an action theoretical approach, a subject-subject relationship) but each of which also emphasise different aspects.

- Pragmatic action research (elaborated by Greenwood and Levin (1998)), with two central parameters: 'Knowledge generation through action and experimentation, and the role of participatory democracy'(2001, p. 104).
- Co-operative inquiry (in particular Reason and Bradbury, 2001), with strong emphasis on personal growth, self-actualisation and inquiry into personal strength.
- Critical (participatory, emancipatory and exemplary) AR, grounded by critical hermeneutics and often by neo-Marxist theories of sociology, psychology and education.

In this chapter I won't address all similarities and differences between the different approaches. Instead I will take the leading themes of this chapter and find out what lines can be discovered within and between the approaches Boog distinguishes. One issue of terminology I want to address here: Boog places Participatory Action Research (PAR) in the stream of critical AR, in contrast to the use of the term in many studies in the South (see e.g. Borda, 2006), where PAR is defined as the 'umbrella' under which many forms of AR can be found, this latter is the way I use the word AR in this contribution.

11.2 Researching by change, changing by research: who sets the goal?

In chapter one we defined as a characteristic of research in action the strong orientation towards social and emancipatory goals from the perspective of people who are usually deprived of resources. Action research is meant to make undesirable situations such as inequality, poverty, oppression, better. This situation is realised by involving those who are normally in the position of underdog in the research. For the researcher this implies a number of important choices. The first one to address is how to deal with the setting of the goals of the research and who will be involved in making those choices: what is the result aimed for and what process leads to the formulation of this aim?

Important considerations when setting goals for change and making general notions of empowerment, democracy and emancipation more concrete and fitting for specific contexts are partly the result of the relationship between researcher and researched party. When the researcher enters a certain domain for research 'from the outside', he or she has goals stemming from a personal context, political views and personal biography. We see a number of differences in handling this and go into the main features of these differences.

Activist research: the researcher has analyses of society that form the basis of the research. Gemma van der Haar (Chapter 7 of this volume) quotes activist anthropologist Charles Hale who makes a plea for 'making our policies explicit and up-front, reflecting honestly and systematically on how they have shaped our understanding of the problem at hand.' (2001: 14). From this analysis goals are deduced to be realised through the research (for instance, strengthening of the position of employees toward employers, teaching people to cope with HIV/Aids, etc.). With these goals in hand the researcher tries to find partners who share them as the basis for cooperation. In many cases these partners are organised as a collective, such as trade unions, social movements, communities, etc.

Activist research could be placed in Marxist (and sometimes feminist) schools (as part of the critical AR approach), where researcher and researched party share the same ideology. Recently we see also versions of it, where AR is deployed to implement policy. In that case policy is formulated so the goal of the action is already fixed and AR is used as an instrument to increase support. As one of our students analysed: '*In our country action research is seen as politically correct. Action researchers seem to have a lot of possibilities for funding if they follow government policy. The question is how free researchers and co-researchers can remain in defining the problem and seeking adequate solutions.*'

On the other side of the spectrum we have the researcher who listens to the researched party and tries to leave his own position out of the picture. The researched party indicates the goal of change; the researcher helps them to formulate this goal. In this version the researcher's point of departure is to leave his own objective aside to offer optimal space for the researched party to find words for their own objectives. Many interpretative and ethnographic researchers choose this position. It is also observed that the choice for the domain and question of the research is in hands of the researcher and is in this way determined by his or her motivation and goals.

Another and third approach is called the reflexive or reciprocal approach, of which exemplarian action research is one of the designs. Here the researcher explicates his or her own objectives and starts a dialogue with the researched party, as a result of which common objectives for the research are fixed (see for instance Coenen, 1989; Beukema and Valkenburg, Chapter 4 in this volume). The point of departure for the researcher is that it is impossible to leave her own objectives aside in her own actions. In order not to ignore these objectives, because at the same time they play (conscious or unconscious) their role in the choices of the researcher and her actions, explicit formulation is necessary. This explication doesn't mean that the researched party doesn't have a say in the final formulation of objectives of the research: the researcher brings her own objectives into the discussion, at the same time as she is (re)searching the objectives of the researched party/co-researcher. In the following discussion common ground is debated and joint objectives for research are formulated. This process often continues during the whole research project: the research starts with consensus in words, but in practice it turns out that the meaning of these words is different for the various participants in the research. Because of these different meanings continuous explication of one's own actions in relation to the formulated objectives of the research is necessary. At the same time this forms an adequate tool for refining and concretising the fixed goals.

The researcher might also be part of the group aimed at in the research. In practitioners research the researcher is looking at his own individual situation. Researcher and researched party are the same in this case, which means a minimum of discussion on the objective of the research. The discussion is focussed more on the social relevance and the possibility of generalisation of the outcomes. This kind of research can be found in the field of education.

In co-operative inquiry a reciprocal relationship is also the starting point of research. Here the aim is not defined in terms of objectives of research, but as '*a vision of persons in reciprocal relation using the full range of their sensibilities to inquire together any aspect of the human condition with which the transparent body-mind can engage.*' (Heron, 1996: 1) The co-operation is based upon similar concerns and interests (Heron and Reason, 2001). The aim is to get bootstrap groups to set up their own inquiry and as part of that formulate their own goals. However, this is something people have to learn and is not taken up automatically by these groups. In the current situation researchers will ask others to call research groups into being. Such groups will find it helpful for the initiators to guide them into the method while they gradually make it their own (Heron, 1996: 62).

When the researcher is part of the target group of the research the situation gets more complicated. Here the researcher doesn't necessarily focus on his own situation in particular, but focuses on the situation of the group of which he or she is part. For instance, when feminist researchers are considering the way in which they can match their own goals of the research with those of the women the research is focussing on (e.g. Lennie *et al.*, 2003). In fact, we see issues at stake here that are comparable to what I described about the situation where the researcher comes from outside the researched party. But here (reflection on) the own biography is more crucial.

Finally there is contract research: a client has certain goals for research and is searching for a suitable researcher. This researcher is hired to reach the goals set. When the given goals are not shared, the researcher won't get the order. In these cases the question is to what extent the researcher commits her-/himself, even when the research doesn't fit the own goals entirely. Here also, different paths are taken: those with a pragmatic attitude will step into the research sooner and strive to reach the own goals as best they can on the way. Those with a more principled attitude will want to fix their own scope beforehand. Fricke indicates at this point that action researchers face the task of breaking through conventional images of research, found among circles of management and policy makers: the researcher is either seen as an academic who doesn't get involved in the practice, or as an executive consultant who's coming to fix a practical problem. AR is neither; on the contrary, a central characteristic is that, although the direction of the process can be settled beforehand, the precise outcome can't.

11.3 Research as a cyclical process

The second question about AR is the extent to which research is part of the process of change itself: does the research stop after analyses of the current situation and formulation of recommendations for improving this situation by the researcher, leaving the researched party to decide how to deal with these recommendations? This question is answered negatively by most action researchers: research continues while the process of change is taking shape. Action researchers usually see their research as a cyclical process. Unlike the empiric-analytical tradition, in AR research and acting are not separated (see e.g. Coenen, 1989: 88 ff) In the Netherlands this is often referred to as the difference between the empiric-analytical cycle of the dominant predictive scholarly work, as explicated by De Groot (1961) and the regulative cycle of Van Strien (1975), 'a methodological design of problem and practice oriented thinking.'(Coenen, 1989: 93). In this regulative cycle the focus is not only on studying but also on regulating social reality: realising set goals or avoiding undesirable situations.

In this cycle directional question for the diagnosis of the practice under research is the meaning for realisation of the situation as desired. The diagnosis offers a design for intervention that is evaluated; evaluation leads to a reformulation of the problem, etc. Traditional empiric-analytical research on the contrary diagnoses the situation under study on the basis of hypothesis derived from relevant theory and makes recommendations for policy makers to apply. In this sense it is theory-driven. AR as a practice-driven approach starts with the practice under study, diagnoses (with the help of dialogue and theory) but also designs the intervention (with the help of practical knowledge of the actors involved and with theory as input of the researcher) and continues research during the implementation of this design to realise a process that leads to intended change.

In essence this cycle can be found in much literature on AR, although as a result of the practice-oriented character of AR the design of a whole series of cycles will hardly ever be undertaken beforehand. (See e.g. Reason and Bradbury, 2001; Greenwood and Levin, 1998, etc.). A central factor is that in addition to diagnosis, the planning of the intervention and the intervention itself is formed by the researcher in cooperation with the co-researchers. In this sense AR isn't limited to application of 'real' scientific research, where methodological knowledge is reserved for the researcher. Because of this cyclical approach the complexity of AR is a given fact. For instance, the cycle is the mainstream of the research but the meaning of different steps can be different for the actors involved; new actors come on to the scene and bring in there own meanings and values. Moreover, the intervention has not only intended, but also unintended consequences of action that have to be analysed and given a role in the process of research and change. It is possible that during the intervention something unforeseen happens which necessitates a redefinition of the problem. It is the competence of action researchers to maintain an overview of the process and set up the evaluation in such a way that a redefinition can be recognised and dealt with by those involved.

So, Figure 1 only looks simple at first sight: we have a step-by-step-plan and let's follow that. Alas, the figure isn't meant to apply reduction of complexity. The practices of action research show different ways of handling this.

Some action researchers follow the given steps as a dynamic process for reaching a fixed goal. The activist researcher as we've met in the previous paragraph has such a set goal. These goals lay down a large part of the cycle, concerning both the problem definition as well as the subsequent steps in the research project including the intervention. The input of the researcher will be rather decisive. Another approach is the one in which the researcher has indeed set the goal, but these goals concern the change process. For instance, Argyris (1976, 1991) in his theory of learning organisations gives AR a central place. Members of organisations shouldn't only learn to evaluate their action, but also to reflect the preconditions that underlie these actions. AR is aimed at learning this while learning: in the process of change the reflexive competences of members of the organisation are increased in such a way that they can be used fully in the new situation. The functional layout and top-down steering of the traditional, fordist organisation is good cause for a narrow-minded view of organisation members that forms an obstacle for learning. For Argyris a key reason to plead for AR is as an instrument for transformation from the old 'Model I' to a new 'Model II' organisation. In these new organisations free but well-founded choices can be made on the basis of valid and verifiable information, where the personal responsibility of its members to test effectiveness is a starting point. Here the goal is not set for the content but preconditions are formulated that are strived for in and through the research. In this way the design for the new situation is partly fixed, as is a large part of the diagnosis. AR is brought into action to contextualise the design of Model II to the organisation concerned.

Another way of dealing with the regulative cycle is as an aid to determining the direction of change. An overall line is fixed and the cycle is used to clarify the different steps in the

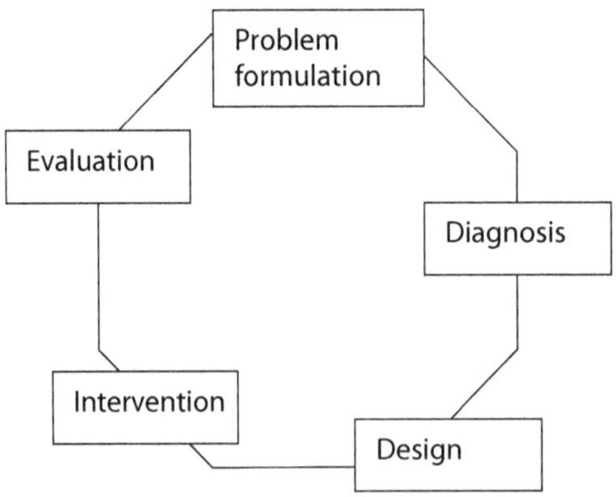

Figure 1. The regulative cycle.

process. Ramaru in this volume gives a refined example of such an approach to the process: the steps themselves, as well as the various constituent processes within these overall lines are distinguished. The cycle is performed within each step in the overall process. This is also the case in co-operative inquiry, where different, but interrelated kinds of knowing are distinguished (presentational, propositional, practical en experimental knowing). These different forms of knowing play different parts in the various stages of the research cycle (Heron, 1996: 52 ff).

In principle the cycle can also be refined in multilevel terms. Where the individual level is included, we can say that each individual has her process of change within the trajectory of the collectivity she is part of. But while changing herself the individual helps structure the collective process and the changes taking place on that level. In other words, the individual produces the change process of the collectivity (organisation, community, etc.) and is also influenced by it. Beukema and Valkenburg in their contribution (Chapter 4 in this book) try to get grips with this multilevel process. In this approach the cycle is an aid to –given the goals of research- clarify the complexity: different levels of change are included as well as different tempi in which actors can be part of this change (Coenen, 1989; Tromp, 2004).

11.4 The relationship between researcher and researched party: between involvement and distance

The relationship between researcher and researched party assumes a central place in the discussion on AR. This becomes clear when looking at the different concepts that are used for the researched party: co-researchers, experts on everyday knowledge, etc. It is with good reason that we speak of a fundamental debate: at this point AR is a challenge for orthodox consensus in science, where the researchers keep as much distance to the domain of research, including the people concerned (see also the Tromp's chapter in this volume).

Action researchers are looking for new ways to define their relationship with the researched party and they do so in different ways. Firstly this concerns the moment of active involvement of the researched party into the research and the place they'll have in the research process. But the nature of the relationship between researcher and researched party is also at stake: how can reality be approached as adequate as possible, how is the recognition of the different kinds of knowledge of parties involved. Both questions apply to the relationship between involvement and distance: how involved can or should a researcher be not only to find words for the situation of the co-researchers, but also to understand it fully. And is distance necessary to analyse the situation under study adequately and if so, how does this distance come about? In fact, at stake is the translation of a new relationship between mode one, two and three knowledge as Tromp (following Novotny) has identified in Chapter 10 in this volume.

Finally, the combination of different kinds of knowledge that are used in AR requires doing justice to the actors who bear this knowledge and are willing to communicate it with others. As a consequence special attention and creativity is required to find a climate of exchange of the existing knowledge and adequate methods for all involved to acquire and enrich it. At this point action researchers add more interactive methods to the more common research methods (interviews, participatory observation, etc.). I will elaborate on these methods in the last part of this paragraph.

The position of the researched party in the research process

When we focus on the involvement of the researched party in the research cycle we can distinguish a continuum from non-involvement at the one end and total involvement at the other. In order to clarify the distinction between AR and other forms of practice-oriented research, I start with a somewhat broader scope than the different approaches of AR as mentioned in the introduction of this contribution.

Action research is sometimes defined as the 'research of actions of others'. However, in my opinion for this form of research it is better to speak of 'intervention research' (for instance, evaluation research, policy research). The researcher isn't involved in the process at all, but is focussing on the interventions of the researched party. She/he is looking from the outside at this intervention, either at the process of it, the outcome of it or a combination of both. One step further towards the concept of AR is where an intervention is planned *and* implemented by researcher and researched party. They co-operate in formulating the aim of the research and implementing the intervention. However, the gathering and analysing of the data as well as formulating conclusions and recommendations are in the hands of the researcher. As an expert on research she/he is studying the process and/or result of the intervention without participation of the researched party. Co-operation continues when it comes to analysing the consequences of the conclusions and recommendations for the situation under study.

Participatory and pragmatic AR takes a further step in the co-operation between researcher and researched party. Intervention and research are activities of both parties; both are involved in setting the aim as well as doing the research as such. In participatory action research there is a strong emphasis on democratic relations between researcher and co-researchers during the research and as a result of it. Learning in communication processes between insiders ('owners' of the problem) and outsiders (professional researchers who seek to facilitate a co-learning process aimed at solving local problems) is seen as a central element of AR (Greenwood and Levin, 1998: 115 ff). The involvement of insiders forms the basis for local usefulness of the results of the research; the involvement of professional researchers opens up possibilities for systematic analysis and a firm basis for interventions. Generalisations from the research are sometimes possible, depending on whether the local results give insights relevant for an audience beyond the co-researchers in a particular project. (Greenwood and Levin, 1998: 122). The outcomes of the research are firstly used in the local context and vice versa: the workability of the results in the local context is an indication of their trustworthiness.

In a co-operative inquiry groups of co-researchers are involved in the research cycle as a whole. Professional researchers are seen as teachers for co-researchers to do their own research. It is best if the involvement of the researchers ends as soon as the co-researchers are able to research their own situation. Examples of co-operative inquiry mainly concern professionals who together through continuing reflection, action, reflection on action etc. try to improve their professional activity. In addition, groups of patients gather to improve their position in relation to medical staff. Researchers support these groups in a methodological way; their position can be an outsider, but sometimes they are also part of the group (see for an overview e.g. Heron and Reason, 2001: 179 ff).

Involvement of co-researchers and researchers on the basis of equivalence is also central in exemplarian or reflexive AR. The competence of the researched party is seen to be of paramount importance, a competence which is not only attributed to the actor but is also present in the actor's self-assessment. (Coenen, 1996: 5) But here the focus is not only on improving the competences of the co-researchers to improve the situation under study. Research is '*first and foremost, empirical research into the actions of others*' (*ibid*.: 5) and in this way seen as knowledge production that can also be made use of in comparable contexts such as the one under study. The local is central, but it is analysed in its context and taken as a starting point for knowledge production and the search for generalisation. This implies another role, not as much for the co-researchers, but for the researchers: not only facilitating the local research process but also (preferably together with the co-researchers) relating this process to the public and scientific discussion on the themes that emerge from the research.

In all these choices it is important to recognise that the 'researched party' do not automatically become 'co-researchers' just because this fits the approach of the researcher. People often have a traditional view of research, where the researcher holds the truth and one's own knowledge of the situation can at best be introduced passively. Boog *et al.* (2005: 175) recommend '*to familiarise participants as soon as possible with their role as co-researcher*'. In addition I would state that a central part of the research process is to explain the role of co-researchers and complement this cooperatively. This complementing can be different in various contexts, but mutual recognition of one another and each role in the research process is crucial. For this reason many AR projects take a relatively long time to start up: in this phase parties have to agree on the goal and design of the research *and* on the internal relationships. In many cases relations between parties involved only become clear during the research process (see the different cases in this volume).

The nature of the relationship between researcher and co-researchers

It is not just cooperation in the different steps of the research process that colours the relationship between researcher and co-researchers. The nature of this co-operation is equally important. Tromp already focussed on this matter in her analysis of the changing role of researcher and researched and of the mutual learning process of both parties.

This theoretical view has far-reaching consequences for the design and implementation of the research. Because of this essential elements of AR include the organisation of the dialogue between researcher and co-researchers, reflection as an ongoing learning process of those involved in the research, etc. Much literature on AR looks into the possibilities and dilemmas of striving for equality in the relationships between different parties in the process: how to deal with differences in knowledge, in position, in involvement. And are the implications of all these differences for the democratic modelling of the research process itself? (see e.g. Gergen, 2003; Lennie *et al.*, 2003)

Methodological instrumentation

The nature of the relationship between researcher and co-researchers demands reflection on the chosen methods of data gathering and analysis. For gathering data action researchers sometimes use 'regular' research methods, such as in-depth-interviews, participatory observation and – if helpful in getting an overview of the situation- sometimes quantitative methods as well. They often make a plea for triangulation. For instance, Kibwika (2006: 36) in his study of innovative competences in African universities uses different methods of data gathering, such as interviewing, workshops, documentary reviews, participant observation and self assessment. (Keune *et al.*, Chapter 2 in this volume discuss not only this triangulation of methods, but also other forms of triangulation).

Next to these kinds of methods we also see action researchers using interactive methods of data gathering and analysis, in which co-researchers take an active part. They are creative in trying to find adequate instruments for the context under study. Within the scope of this contribution it is impossible to pretend to give a complete overview. Which is why I present a few examples to give the reader an overall impression of the possibilities.

With professionals from the social service sector an AR project used the method of 'prototyping' to improve the care plans and the way professionals used them in their everyday practice (Van Boheemen and Schippers, 2005: 75 ff). A prototype is described as '*a working model of (eventually parts of) a system, that emphasises certain aspects.*' (p.75). By testing the prototype of a care plan in everyday practice by the cycle of action research, researchers and co-researchers could jointly collect and analyse information. This led to a continuous readjustment of the prototype until the design of the plan satisfied all those involved.

A search conference is a research instrument that is often used in situations of group learning (Martin, 2001: 200 ff; Greenwood and Levin, 1998: 155 ff). Martin, following Bunker and Alban, defines these as '*events designed to engage representatives of an entire system whether it be an organisation or a community, in thinking through and planning change.*' (2001: 200) Search conferences are meetings (mostly of several days), in which researcher and co-researchers through discussion, simulation games and other learning techniques cooperatively search for an adequate analysis of the situation at hand and the possibilities of change. Search conferences can focus on diagnosis and the design of action plans. They can also be organised later in the process to evaluate the interventions and analyse them for their consequences. Greenwood and Levin developed a conference design for the starting phase of the research process in which the following steps are distinguished: Review of past to present, review of present to future, problem identification, possible lines of actions from now to future (1998: 163).

In the contributions in this volume by Ramaru (workshops, Chapter 3) and Beukema and Valkenburg (training and intervision, Chapter 4) examples of search conferences are described.

It is essential in action research for the co-researchers to have a say in the adequacy of the results of the research. This implies that – even when the researcher is the one who gathers data and forms a picture of the situation- this picture is presented to the co-researchers. The question as to whether this picture is recognisable and complete often forms the start of the research – learning process. Co-researchers are not used to being approached in this way by researchers. They learn to participate actively in the research process and in this way to reflect on their own situation. The researcher learns by communicating with co-researchers the significance of the data gathered and analysed in the specific context of the research. This brings us to the last topic of this contribution: the mutual adequacy in AR as central element in determining the validity of the results.

11.5 Mutual adequacy as parameter of validity in action research

Because of the difference in the relationship between researcher and co-researchers as compared to traditional research, much debate on AR is on its reliability and validity. From the perspective of more traditional researchers (empirical analytical as well as interpretative) it is inconceivable that scientific knowledge can be produced in cooperation between scientists and non-scientists. From the orthodox point of view, objective and scientifically adequate knowledge is knowledge in which the subjectivity of the researcher and of the people (s)he is researching, is kept out of the research situation as much as possible. This refers to another assumption inherent in the orthodox consensus, i.e. that there is a fundamental distinction to be made between scientific and common-sense knowledge. Consequently, strictly separated roles are defined for both parties in the research process. (Valkenburg and Lind, 2002). This perspective forms the basis of rules of reliability and validity. It is clear that action research can not do much with these rules. For instance, reliability in the orthodox consensus is often defined as finding the same results when repeating the research. But while AR strives for change, repetition is not an issue.

For AR this means that the validity of results has to be developed along other lines. As one of the leading scholars on co-operative inquiry Heron states, '*The challenge after positivism is to redefine truth and validity in ways that honour the generative, creative role of the human mind in all forms of knowing.*' (1996: 13) In all approaches of AR as described in this contribution validity is a central point of discussion. In all approaches triangulation is seen as one of the main ways of validating research results.

In pragmatic action research trustworthiness is an important concept within this framework. In this approach validity and reliability are the result of a discursive process between participants and researchers who negotiate their meaning of the experiences of their action. (Greenwood and Levin, 1996: 114). The logic of action is guiding, if the intended results follow from the action, the research has proved to be worthwhile.

In cooperative inquiry a large number of procedures are developed to increase validity. These procedures, together with inquiry skills, aim to free various forms of knowing involved from some of the distortions of uncritical subjectivity (Heron, 1996: 131). These procedures are related to the various phases of the research cycle. For instance, by combining individual and collective reflection on actions undertaken side circles is constructed to jointly validate the generated knowledge. Another important starting point for validation is using 'polarity of method' (for instance divergence – convergence polarity, individual – group polarity, heterogeneity – homogeneity polarity, etc.) (Heron, 1996: 138). By combining these polarities in scheme (co-) researchers can get an overview of the processes taking place in the research and discussing these processes. For an overview of validity procedures in cooperative inquiry, readers are referred to further discussion on this point by Heron (1996).

In exemplarian action research the central argument in the discussion about quality and validity is the double hermeneutics, as stressed by Giddens (1984) among others. The researchers are interpreting a reality that has already been interpreted by the people under study as competent actors. These actors use their knowledge of themselves and their social context to translate their motivations and intentions into actions and at the same time monitor this process in a reflexive way. Their knowledge is not 'just' about facts, but also about theories and concepts. As in other forms of action research, social sciences are seen as forms of intervention that aim at social change and the empowerment of actors. More specifically, exemplary action research is based on a theory of action in which social reality (and thus social change) is the result of contextual, reflexive interactions. Social structures are at the same time conditions for and results of these interactions. To understand social reality and the possibilities for social change, knowledge of the specific context and the underlying motives, plans, insecurities, routines and competences of the actors involved is essential. Therefore, to understand social reality is to understand the reflexivity of the actors involved and the way they reflexively structure their everyday reality. This means that the quality and validity of research results is primarily, though not exclusively, determined by the relationship between the researcher and the co-researchers. Beukema and Valkenburg formulate starting points of the aim to achieve quality and validity in exemplarian action as follows (2007: 171-172):

- understanding social reality as the result of contextual, reflexive interactions of competent actors;
- developing possibilities for changing social reality;
- supporting and empowering the actors concerned to put these opportunities into practice.

This, in turn, requires that research should be based on the principal of reciprocal adequacy and a relationship of trust between researchers and co-researchers.

11.6 To conclude

Writing an overview as I have done in this contribution is a succession of making choices about what might be important for the reader who is starting an AR her/himself. This means inevitably that interesting and useful thoughts of fellow researchers are left out. When action researchers state that their research cannot be value-free, the same can be said of my stand in the debate on AR. As an exemplarian action researcher I have tried to do right by other approaches as best as I could. Nevertheless, due to the time available and set priorities, mutual adequacy was not possible: I've written about other scholars without asking them if they could share my interpretations. That's why I see this chapter not only as contribution to this volume but also to further discussion on the future of action research.

References

Argyris, C., 1996. Leren in en door organisaties. Het hanteerbaar maken van kennis. Scriptum Books, Schiedam.

Beukema, L. and B. Valkenburg, 2007. Demand-driven care in the Netherlands. A case of exemplarian action research. Action Research 5: 161-181.

Boog, B., 2003. The emancipatory character of action research, its history and the present state of the art. Journal of Community and Applied Social Psychology 13: 426-438.

Boog, B., M. Slagter, I. Jacobs-Moonen and F. Meijering, 2005. Synopsis: handreiking handelingsonderzoek. In: B. Boog, M. Slagter, I. Jacobs-Moonen and F. Meijering (ed.), Focus op action research. De professional als handelingsonderzoeker. Van Gorcum, Assen, pp. 154-186.

Borda, O.F., 2006. The North-South convergence: A 30-year first-person assessment of PAR. Action Research 4: 351-358.

Coenen, H., 1989. Handelingsonderzoek als exemplarisch leren. Van Arkel, Utrecht.

Coenen, H., 1996. Towards a closer definition of action research. In: B. Boog, H. Coenen, A.W.M. Keune and R. Lammerts (eds.), Theory and Practice of action research. With special reference to the Netherlands. Tilburg University Press, Tilburg, pp. 1-13.

De Groot, A.D., 1961. Methodologie. Den Haag.

Gergen, K.J., 2003. Action research and orders of democracy. Action Research 1: 39-56.

Giddens, A., 1984. The constitution of society. Polity Press, Cambridge.

Greenwood, D.J. and M. Levin, 1998. Introduction to action research. Social research for social change. Sage, Thousand Oaks, London, New Delhi.

Hale, C.R., 2001. What is activist research? SSRC 2 (1-2): 13-15.

Heron, J., 1996. Co-operative inquiry. Research into the human condition. Sage, Thousand Oaks, London, New Delhi.

Heron, J. and P. Reason, 2001. The practice of co-operative inquiry: Research 'with' rather than 'on' people. In: Reason, P. and Bradbury, H., Handbook of action research. Sage, Thousand Oaks, London, New Delhi, pp. 179-188.

Kibwika, P., 2006. Learning to make change. Developing innovation competence for recreating the African university of the 21st century. Wageningen Academic Publishers, Wageningen.

Lennie, J., C. Hatcher and W. Morgan, 2003. Feminist discourses of (dis)empowerment in an action research project involving rural women and communication technologies. Action Research 1: 57-80.

Martin, A.W., 2001. Large-group processes as action research. In: Reason, P. and Bradbury, H. (ed) Handbook of action research. Sage, Thousand Oaks, London, New Delhi, pp. 200-208.

Reason, P. and H. Bradbury, 2001. Handbook of action research. Sage, Thousand Oaks, London, New Delhi.

Tromp, C., 2004. Wide-angle science. A critical-reflexive research model based on a broad rationality concept. Jan van Arkel, Utrecht.

Valkenburg, B. and J. Lind, 2002. Orthodoxy and reflexivity in international comparative analysis. In: van Berkel, R. and Hornemann Moller, I., Active social policies in the EU. Policy Press, Bristol.

Van Boheemen, M. and A. Schippers, 2005. The method is the message. In: Boog, B., Slagter, M., Jacobs-Moonen, I. and Meijering, F. (ed.) Focus op action research. De professional als handelingsonderzoeker. Van Gorcum, Assen, pp. 75-84.

Van Strien, P.J., 1975. Naar een methodologie van het praktijkdenken in de sociale wetenschappen. Nederlands Tijdschrift voor de Psychologie 30: 601-620.

12. Research in action

The core business

Coyan Tromp, Leni Beukema and Conny Almekinders

At the end of this book, we fulfil our promise to address a number of core questions of research in action. In dealing with these core issues in various ways and illustrating them in different forms, the various contributions in this book have been an important input in the formulation of our response.

12.1 What does research in action stand for?

A dynamic research strategy and scientific paradigm

The general aim of the book has been to bring to light a type of research and paradigm that acknowledges and represents the high quality standards of scientific knowledge creation, without losing touch with the actual needs that exist in society. An approach, moreover, that explicitly focuses on the question of how scientific knowledge can be put to use and help improve the situation for people in concrete practices on the one hand, and how this scientific knowledge can be developed with these practices as a firm base on the other. As indicated in Chapter 1, for some, the approach of research in action primarily means an addition to the research strategies they are already familiar with. Others, who are also interested in the wider methodological implications, may view it as an alternative outlook on science.

These different views on the approach are also present in the contributions to this book. However, what they have in common is that they all propose a research approach in which the academic domain does not remain an isolated island, but reaches out to the societal domains – be it public service health, agriculture or welfare work or any other domain – and involves the stakeholders in the research process. Conceptual frameworks from various disciplines are combined to formulate a commonly shared problem definition. For this is another characteristic of the approach proposed here: the boundaries within and between scientific disciplines are not taken for granted. It is the formulated question at hand that is taken as a starting point to organise the necessary input. Sometimes, even the research question or problem definition themselves are not solely determined by scientists. Instead, they are formulated in a mutual interactive and iterative consultancy between all the parties that take part in the research.

When the stakeholders from outside the scientific domain are also involved in formulating the problem, the interdisciplinary approach becomes a transdisciplinary approach. The sources of knowledge relied on are not restricted to the scientific domain either. Non-formal or common sense knowledge, like knowledge based on practical experience and knowledge not related to any particular scientific discipline, is also being used in the knowledge creation process. The perspectives of the various stakeholders are taken into consideration and their knowledge and expertise is welcomed and blended with the available scientific knowledge (see Keune *et al.*, Chapter 2; Ramaru *et al.*, Chapter 3 and Beukema and Valkenburg, Chapter 4).

After positioning the approach in the field of science the next step is to develop a suitable research design in which a range of methods can be employed. Those who view research in action not merely as one more research strategy, but as an alternative paradigm, argue that it forms more than just a toolbox of research methods. They (as we do) propose to view a methodology as a framework for analysis that contains the rules about how (a specific form of) research is to be done and how the various methods are employed. The conception of a research approach as a paradigm is based on the consideration that we hold certain theoretical presumptions that form our perspectives, direct our observations and colour our interpretations. 'Vision' is never a neutral act; it is prompted by basic theoretical and philosophical ideas (see the Chapters 9 and 10 of Verschoor and Tromp in part II of the book). Thus, no explanatory model is to be taken for granted. Therefore, researchers should be explicit about which vision is underlying their methodology and about which rules they follow and why. Another implication is that they should in principle be able to explain why they think certain research methods are suited for specific research and others are not. Moreover, this methodological stance leads to alternative conceptions of scientific values and criteria like objectivity, validity, adequacy and generalisability, as we will see in the remainder of this chapter.

The incorporation of various kinds of knowledge

Traditionally, universities concentrate upon the production of mode 1 knowledge, i.e. cognitive-instrumental knowledge aimed to bring about objective, value-free and universally valid knowledge concerning a specific scientific subject field (Gibbons *et al.*, 1994, Tromp, Chapter 10 in this volume). This is a highly valued type of knowledge and rightfully so, since it is an important potential source for both technological and societal innovation as well as for social change. But this type of knowledge has its limitations and can only to a certain extent be expected to be useful outside the controlled environment of academic research laboratories. The production of this type of knowledge relies solely upon the scientific researchers, who extract the needed information from their data sources in or outside the academy.

In the approach presented here, other types of knowledge are considered important as well. Especially when we are searching for knowledge that can help enhance innovation and social change, we need a type of knowledge creation that takes into account both the expert knowledge of scientists and the particular expert knowledge of the other stakeholders.

The knowledge of users usually refers to the direct context in which the new knowledge will be applied, e.g. the knowledge about the functioning of the hospital in which a new schedule is planned, the farm where a new tool is being used. Since scientific knowledge increasingly plays a role in the thoughts, actions and practices of people in the developed world (see Valkenburg *et al.*, Chapter 1) but also in the developing countries, this contextual knowledge of users reaches further than the concrete context itself. Thus, their input implies more than 'just' knowledge of their own particular context – which taken by itself would already contain an important added value. Notwithstanding this broader impact, this way of knowledge development leads to research results that are in first instance 'applied' of character and highly contextualised. We have come to know this type of knowledge as mode 2 knowledge: practical knowledge, focused on finding scientifically legitimated solutions for practical questions that are tied up with economic aspects and policy matters (Gibbons *et al.*, 1994, see Tromp, Chapter 10 in this book). Within the mode 2 domain, the external validation of the knowledge that is developed is not always a matter of great concern – a fact that is reflected in the absence of this issue in most of the contributions. The main question is not whether the created knowledge can be generalised to other situations, but whether it can be put to use within this specific context. While mode 2 knowledge is typically judged by its practical applicability, we mustn't forget that it always builds upon the results of the production of mode 1 knowledge (and vice versa). And as we know, the question of generalisability and external validation is regarded as highly important within the domain of mode 1 (see also Tromp, 2008).

The ultimate goal is to develop knowledge that is relevant for the solution of complex societal problems. This means, that the knowledge generated in specific circumstances needs to be translated to a higher level of abstraction, to be able to transfer it to other, similar situations. Since in the context-bound knowledge, structural factors are involved, to some extent generalisation must be possible. To find out whether the knowledge that was created in this particular situation can also be applied to and used in another situation, we have to look into the complexity and contextuality on various levels, like the functioning of the system that was studied, the composition and background of the social practices that are at stake, the positions of relevant (f)actors, the relation between relevant (f)actors and to similarities and differences in the institutional set-up of a field.

In short, we try to transcend the context-bound knowledge by looking for similarities and analogies in natural systems and social contexts (exemplars and family resemblances) to aim for some degree of generalisability. The issue is the comparability of various contexts and situations; what resemblances are there, which differences can be observed? In the end, it is up to the community of scientific researchers and co-researchers to judge the comparability of contexts and the actors in those contexts. This kind of generalisation suits the aims connected to mode 2 knowledge creation and is of a different kind than the one within the positivist conception. Within positivist research the focus is on the search for invariant laws or on the application of general causal structures to a specific case (induction or deduction, or a combination of the two: abduction). This outlook on generalisability often based on statistical reasoning, is most suitable for mode 1 knowledge production. Though generalisability forms part of the comparative reasoning ('are those causal mechanisms that we have found in a controlled environment also effective in the open system we have been studying?'), it is only one of the aspects we reckon with in knowledge generation. The comparison is not just made from a systemic perspective, but rather from the perspective of the interrelatedness of (both natural and social) systems, structures and actors. This means that, besides cause – effect relationships, we also should reckon with actor causality, referring to a relationship between intentions and behaviour/actions. The latter type of causality (actor causality) can cause the former (event causality) not to emerge, or emerge in a different way than we would have expected from what we learned from experiments under controlled conditions.

We also need a third type of knowledge; intuitive knowledge which is focused on the elaboration of existential and moral insights in view of practical questions and dilemmas. This mode 3 knowledge creation does not usually follow a linear line, but rather has its own logic and is shaped by a certain friction between different points of view. The learning processes needed to bring to the fore mode 3 knowledge can feed upon theoretical explanations and practical deliberations, but also calls for the input of inspiring narratives, visions and metaphors with regard to moral dilemmas and the big questions in life (Tromp, Chapter 10 in this volume; Kunneman, 2005: 116 ff).

This range of types of knowledge from 1 to 2 and 3 shows that the proposed research model has its starting point in a more advanced conception of rationality. By not only considering mode 1 knowledge, but integrating mode 2 and 3 knowledge as well, this rationality concept offers a wide-angle perspective on reality that enables us to take into consideration not only factual but also moral, ethical, political and power aspects that are related to the complex issues were are studying.

A comprehensive framework for evaluation

We have also seen that the different kinds of knowledge lead to various regimes of justification (see Verschoor's first contribution to this book). The cognitive-instrumental rationality implies a view of means-ends relations that is quite confined, compared to the ways in which means-ends relations are viewed from a broader perspective of rationality. For, besides a purely goal-oriented, economic view of what scientific research should contribute to, other views are possible too. When research transcends mere factual matters (which, by the way are never 'just facts' but always imply certain norms, at least selection criteria which lead to inclusion of some, exclusion of other aspects or people and thereby relate to power relations too) and also involves ethical, moral and political issues, the cognitive-instrumental perspective of means-ends will not suffice. We cannot solely rely on justifications that exclusively look at matters from the perspective of efficiency or from an economical point of view (i.e. justifications from the industrial or economical regime). We also want to listen to arguments that emphasise collective welfare (the civic regime) or the importance of local history, culture and tradition (the domestic regime) and stress the environmental effects of technologies and innovations (the 'green' regime). We need, in summary, an advanced evaluative framework and multi-criteria analyses to come to well thought-out solutions to the challenges at hand and to be able to formulate long term, visionary strategies for the future (cf. also Keune *et al.*'s contribution in Chapter 2).

Furthermore, we have claimed that a methodology that takes into account mode 1, mode 2 and mode 3 knowledge as well as the various regimes of justifications that actors rely on to back up their arguments, implies a principal equivalence between the knowledge and competences of researchers and those of the researched (cf. Tromp, Chapter 10 in this volume). This means they are considered as having equal value for mankind. This equivalence has found its translation in the active role that is assigned to both parties within the research process. The practical knowledge of the users and other stakeholders, i.e. their knowledge from experiential learning and sometimes even theoretical knowledge of the particular context are incorporated in the knowledge creation process. And their assessments of the research findings and judgment of the proposed explanations and solutions are seriously taken into account.

What does it stand for?

Research in action describes a dynamic research approach/strategy/paradigm,

- in which both researcher and researched have a voice in the input into and output of the research process,
- which aims to generate and integrate various types of knowledge,
- which is rooted in concrete practices to solve problems in specific situations, but can nevertheless have context-transcending meaning and explanatory power,
- which makes use of and develops knowledge and competencies that are regarded as equal in value for mankind,
- which relies on arguments from various regimes of justification to come to well-considered decisions about complex issues that involve not only factual aspects, but also moral, ethical, political and power aspects,
- with the aim to help find answers and solutions for the challenges our society is faced with and develop strategies for the future.

12.2 For whom, with whom, of whom? And what for?

Various stakeholder groups

We can make a distinction between four parties within research (1) the researcher or the research team; (2) the researched subjects or co-researchers[67]; (3) those who should benefit from the research, i.e. the clients or users, which can be viewed as the critical reference group; and (4) others stakeholders who stand to benefit from the research, for instance policymakers, managers or citizens in general.

To find an answer to the question: 'For whom?', we should define the ultimate stakeholders. To identify this group, the researcher should pose the question as to whose positions need to be improved and whose resources need to be strengthened. A first reaction could be: in principle everyone, for of course we would like to contribute to improvement and positive developments for the whole world, for all parties (cf. Verschoor in Chapter 9 of this volume). But the ambition of research in action can be explicitly emancipatory, i.e. that it aims to specifically improve the situation of those in a precarious position. The main aim of the generated scientific knowledge is then to use it as a resource for the empowerment of the primary stakeholders.

Whether it is really possible to bring about social change and give the researched more power to steer their situation in the direction of improvement depends however to a large extent upon the character of the specific research context. Sometimes, it might be possible to involve the primary stakeholders as co-researchers, as in the case of experimentation with the soil fertility management in South Africa (Ramaru, Chapter 3 in this volume). In that case group 2 and 3 in the categories mentioned above imply the same persons. But often, there is not an option to directly involve the group of primary stakeholders in the research. The researcher then has to deal with their representatives or delegates, like the practitioners working with the primary stakeholders. In that case, you could say that the delegates (i.e. the practitioners, the secondary stakeholders) are the direct co-researchers of the researchers, while the clients or users (the primary stakeholders) are the indirect co-researchers. And this could even go one step further, as for instance the managers (the tertiary stakeholders) of the practitioners (the secondary stakeholders) become the direct co-researchers of the scientific researcher. Of course, there is a limit to representation instead of direct participation of the primary stakeholder group (see Chapter 7 of Van der Haar and Chapter 6 of Roose and De Bie).

[67] Which term is used here will depend on the extent to which the researched subjects are or are not actively participating and collaborating in the research.

The relationship researcher, co-researchers and other parties

The context in which the research relationship is established has major consequences for the relationship between the researcher and the researched as well as for the possible positions they can take. Which organisation or institution put down the request for research and is funding the project? Is it contract research, i.e. conducted at the request of the subsidising government or the management of a business company or a political organisation, or is it initiated by the researcher or by practitioners themselves? How does this influence the relationship between the researchers and the researched? Can the research team maintain an independent position vis-à-vis the research objectives and the primary and secondary stakeholder groups? Are enough resources made available to allow the primary (and secondary) stakeholders to participate in a direct way, or at least a reasonable number of them? These are all factors that influence the character of the research. In short, the question: 'Of whom?' pertains to the issue of who actually 'owns' the problem, i.e. not only who has the responsibility, but also who has the power to formulate what the essential problem is and which strategies are to be followed to change the situation for the better.

We could say that in the most beneficial case, the researchers and the users and/or practitioners support each other. The researchers help to solve problems arising in practice, for which they need the knowledge and the help of the practitioners and users themselves. But when the research is part of a specific mission or assignment of which the objectives are laid down beforehand, it can be questioned whether such a mutually beneficial relationship can be established. It could be that a (local) government or management is sincere in its intentions and wants to profile itself by developing a policy that is scientifically legitimated. But maybe the research merely serves as window dressing. The participation of the users could merely be a means to attach a political correct label to a research project that is basically completely pre-designed and steered by the scientists, the politicians and/or the management. And even when there actually is participation, it might be that important key figures are not included in the selection of the co-researchers. Or that the participation is limited to a single phase and is only used as a means to smooth over the implementation of the proposed changes. Such superfluous involvement of stakeholders never extends beyond mere 'token participation'.

To find out what the situation is, the first thing to do is to check out the goals of the research. What is the research project meant to bring about? In what way do these aims entail an improvement of the situation of the researched, i.e. the primary stakeholders and/or co-researchers? And if they don't, how can they be reformulated so as to bring about these improvements, according to the researchers' perspective? Are these the same goals as those the stakeholders formulate for themselves? If not, how do the goals of the party issuing the research, the researcher and the researched relate to each other? And how can we as researchers handle this relation? Each specific research context will lead to different answers to these questions.

Last but not least we should remember that the researcher or the research team are also people who belong to the answer to the question: 'With whom?' Researchers have their own peculiarities; their involvement in the research brings in individual expertise and strengths, but inevitably also certain flaws. When in the luxurious position of being involved in a research team, researchers have the possibility to put the competencies to optimal use by dividing the tasks in such a way as to profit most from everybody's strengths (and suffer least from unfavourable deficits). Besides methodological competencies the researchers need to have the social skills to be able to deal with the complex relations in the research process (see Almekinders *et al.*, Chapter 5 in this volume). Importantly, the researcher has to be able to bring about a good balance in the various goals that need to be taken into consideration: a modest and to some extent detached attitude should be combined with a strong involvement with the problem situation at hand. But especially in a situation of struggle and conflict, it can be difficult to find a balance between the unconditional commitment that might be demanded by (some of) the parties involved and the open perspective that the researcher wants to maintain (see Van der Haar, Chapter 7 in this volume).

Emancipation and empowerment

Researchers who regard the emancipatory goals as a priority in research in action will strive for an optimal democratic character of the research process (see Chapter 6 of Roose and De Bie and Chapter 2 of Keune *et al.* in this volume). It is not the researcher that decides what the central question of the research should be and how the implementation of change can be realised. It is a democratic research process in which not only the researcher but also policymakers, managers, practitioners and clients have a say in the problem definition and in the change process. Or at least they should have (see Verschoor's contribution in Chapter 9 of the book). Specifically in a situation where the goal is to get from scientific theorising and assessment to decision making, the main questions are: Who is responsible for further research based on the preliminary results? Who should steer the process? Whose knowledge or jugdement is relevant? Who should decide on policy options? Who should be informed about the developments and outcomes during the process? What are the important points of interest during the process?

To complicate things, real involvement of the (primary) stakeholders in the research process does not automatically lead to more social equality in terms of participation and sharing of benefits. For instance, when the researcher's main interest is to generate scientific, theoretical knowledge, it is probably the academic community that profits most, at least more than the actors involved in the research. Moreover, there can also be reasons for one of the involved actors to refrain from active involvement, for instance a lack of trust or reasons of privacy or restrictions in available time and resources. Most of the times, stakeholders cannot and do not have to be active and empowered in every aspect but can move from a passive and detached attitude to active involvement at different times. What is essential is that the choice about (non-)participation must be made on good grounds and must not be the result of manipulation or an evident shortage of time and resources reserved for the research project.

Besides a mutual commitment to the central problem and research question, another

> **For whom, with whom, of whom? And what for?**
>
> Research in action stands for a type of research that is
> - in principle for everyone, since the aim is to stimulate progress and well-being in society in general
> - specifically meant to improve the situation of those in a precarious position, whose voices often are not heard
> - with a researcher or research team, co-researchers and other stakeholders that all bring in their knowledge and competencies in such a way as to benefit the resource process most
> - with the goal of generating new insights and knowledge, both scientific and practical
> - and with the goal of innovating and bringing about social change in order to improve the situation of the primary group of stakeholders.

essential feature of emancipatory research is that it takes as its starting point a subject-subject relationship between researchers and the other actors participating in the research. In the next paragraph, we investigate how we can envisage such a relationship.

12.3 How can we bring about innovation and social change through research?

Research as a mutual learning process

Technological innovation is seldom a matter of 'just' transferring scientific knowledge from the academic domain to the industrial or agricultural or social or services domains. Neither can societal innovation and social change be realised by merely producing new interpretations or by social engineering. We have therefore taken the position that if we want to realise innovation and bring about social change, we have to set up a research project as a mutual learning process. We have stressed the dynamic dimension of knowledge creation; it is a result of the collaboration of scientific researchers, community members and other stakeholders. Contrary to the traditional research approach, research in action is research *with* and not just *about* practice. The researched are not just treated as mere sources of information, but are actively involved as co-researchers – to the extent that this is realistically feasible. This interaction forms part of a multi-layered learning process.

Of course, it is important to formulate explicit goals for a research project. But we have to be very careful that these goals always remain the goals of the participating actors, and not emancipatory objectives prescribed by the researchers, the managers or policy makers that are to lead, through a 'democratic' research process, to a process of innovation and social change. Otherwise it would still be some kind of manipulation or social engineering that does not stand much chance of success (see Chapter 6 of Roose and De Bie and Chapter 7 of Van der Haar).

A (real) democratic research process by definition implies an open research process to which various actors make their own contribution. If there is any emancipation to take place, it is brought about by people themselves, through their participation and cooperation within the research context. Researchers (if not part of the group of stakeholders themselves as, for instance, in some forms of co-operative inquiry) have to dare to remain at a certain distance, to wait and see what happens, to provide room for the stakeholders to critically review their situation (problematisation) in their own words and under their own conditions. The researchers should refrain from the tendency to immediately start looking for solutions (to the problem as they have analysed it) on their own.

One way to dissipate the tension generated by the different positions and roles of researchers is to split up the research tasks and - if there are several researchers - distribute the roles: one researcher could be emphatically present in the field (insider position), whilst the other researcher could take a more distant, coordinating role (outsider position). And if there is a project leader, this researcher can be positioned in between an insider and an outsider position (see Chapter 6 of Roose and De Bie).

The scientific researcher can be expected to have the specific expertise needed to set up the research design and formulate a deliberate and explicit proposal on which methods and techniques to be used. Nevertheless, the co-researchers preferably assist them in this task by using their knowledge of the context for fine-tuning those methods and techniques to suit the actual research situation. For instance, they can check and help to improve on questionnaires or topic lists for interviews and thus play an important role in gathering data. Moreover, they can offer support by talking to key figures in the research field, help with surveys and interviews, do experiments and take measurements. The same applies to the interpretation of data. Thus, one reason to involve the researched as co-researchers is because they carry subtle knowledge about the research situation and have a lot of experience with it (see Chapters 3, 7, 10 and 11 of Ramaru, Van der Haar, Tromp and Beukema). Another, equally important reason is that, since in research data interpretations of an already interpreted reality are implied, the researchers need the co-researchers to check their interpretations (of the interpretations of the researched/co-researchers). To bring about mutual understanding, researchers and co-researchers engage in what we call a double hermeneutic process: the researchers interpret an already interpreted world, the co-researchers comment on these interpretations, the researchers take the comments into account and come up with new interpretations. This process continues during the whole research project. Though the researcher carries the overall responsibility for the research process and its report(s), the results are dealt with in joint consultation between researchers and co-researchers.

Setting up research as a mutual learning process in which the knowledge of both researcher and the co-researchers are regarded as indispensable and complementary in the research process also means that their assessments and validations of the produced knowledge have to be taken seriously. The researchers put their validity claims to the test in critical discussion, not just with their colleague scientists, but also with those involved in the actual research project (see the Chapters 4, 10 and 11 of Beukema and Valkenburg, Tromp and Beukema). For whenever researchers aim to produce practical policy relevant knowledge, they take part in a hermeneutic creation process, i.e. a process in which meanings are given to the phenomena under investigation and to the solutions that are put forward. Since stakeholders with different positions will have different perspectives on the problems and the possible solutions, there will be no 'objective' knowledge about the issue. Therefore, the findings should be double-checked, i.e. agreed upon by both researchers and co-researchers.

In short, in the process from data-interpretation and assessment to decision making, the subject-subject relationship between the researchers and the researched subjects requires that special attention is paid to the collaboration agreements. For the quality of this relationship to a large extent is determining the quality of the results (see the Chapters 2, 5 and 6 of Keune *et al.*, Almekinders *et al.* and Roose and De Bie).

The development of trust as a condition for mutual knowledge creation

To be able to engage in a mutual learning process, there needs to be a basis of trust between the researchers and the researched (this is clear in all contributions, but see specifically Chapters 7, 5 11 of Van der Haar, Almekinders *et al.* and Beukema). This takes time and effort, but is a necessary precondition without which a research project will not be successful. A first step in building trust is to change the general conceptions about the position of a researcher, as well as the conventional ideas about the role of the researched. The latter party gradually needs to begin understanding that the researcher is not the one who will be solely responsible for the problem definition and formulating the research question, but that this is something which has to be discovered during the joint learning process. The same applies for the formulation of the results and the conclusions of the research. Specifically when actors have had prior experience with 'involvement' in research projects, they can be highly suspicious of the researcher. They might want to see proof that the researcher is going to deliver the promised results and might hold back in their participation because they feel they are only considered as sources of information of which knowledge can be extracted without getting anything in return. It is only when they actually see and experience that they are involved in a mutual learning process that they will start trusting the researchers. This is a precondition for growing from a passive role of researched object into a more active role as participant co-researcher.

Participation, collaboration and reciprocal adequacy

To achieve the commitment and involvement necessary for a mutual learning process to take place, participation and collaboration of the co-researchers is an important factor. We have seen that the co-researchers can be given a role in the gathering of data and help formulate the questions and problem definitions. But their participation does not stop there. As mentioned earlier, they also play an important role in the assessment of the validity of the findings and the relevance of the proposed solutions. This part of the research process is no longer considered to be exclusively the domain of the researchers and their academic colleagues. Besides the scientific community other stakeholders are also given a voice in the decision-making process (see the Chapters 2, 4, 10, 11 of Keune *et al.*, Beukema and Valkenburg, Tromp and Beukema).

There are various ways in which the collaboration of the co-researchers in this part of the process can be given shape. They can form a team together with the academically schooled researchers and agree on a certain division in the tasks both parties will perform. Or they can be asked to take a seat on an advisory board or a jury in which managers, politicians, clients and (other) citizens together with scientific experts will judge the reported data and produced knowledge before passing their conclusions and advice to relevant parties.

We realise though, that the objectives and values of the participatory discourse and participatory techniques are based on a quite idealistic image (see Chapters 6 and 7 of Roose and De Bie, and Van der Haar). We should always bear in mind that some actors are involved in the discussion, while others are not. Some people may never be heard, and some discussions never held. In the idealistic image the discussion is open to everyone and nobody is prevented from saying what is on their mind. Both researchers and co-researchers are considered to be autonomous and competent actors who are able to articulate their points of view and opinions in a clear and coherent way, while projecting and accepting critical remarks with regard to the statements that are made during the debate. And that they are able to adopt an open attitude, even when those who have power over them are participating in the discussions. This is an assumption that cannot be taken for granted. Especially when working with people in precarious positions, we should be careful not to consider participation in the ongoing discussions simply as a principle that works because it was stated as a rule. We cannot just assume that researchers and co-researchers are equal parties in the debate. Rather, participation should be regarded as a goal that needs to be worked hard on to be able to attain communicative symmetry in the discussions.

In the end, the knowledge that was created is only considered to be adequate when it is recognised and preferably also accepted as such by both researchers and co-researchers. This principle, which lies at the heart of the research paradigm presented here, is called the principle of reciprocal adequacy. It forms an element of internal validity in the research and is a crucial methodological principle to guarantee the reliability of the data interpretation, forms the foundation for the conception of objectivity we are proposing.

Participation alone does not guarantee that the input of the co-researchers is actually taken seriously; reciprocity is equally important and maybe even more. We could claim that the research objectives – when they are related to empowerment - in certain circumstances can also be realised within a research process in which the researched themselves are not participating directly. For it may be that it is agreed upon by both parties – researchers and researched – that the limited availability of time and resources requires an indirect representation of certain groups of stakeholders. However, the other way around, i.e. participation without reciprocity, does not work. We would face the risk that the participation only serves as an alibi for extractive research that is presented as emancipatory research, but actually asks more of the researched than it returns.

Reciprocity implies a relationship and exchange between the researchers and co-researchers characterised by open communication, mutual trust and respect. How these can best be shaped strongly depends on the specific subject and context and therefore will be different in each research. The relation between empowerment, participation and reciprocity thus remains a complex issue. Just as it is a matter of finding a good balance between the empowerment of people while using (and not abusing) their knowledge, this also implies a negotiation process in which power aspects play an important role.

Therefore, we need to develop innovative methods and develop the social skills that enhance an open attitude from researchers and enable them to reach out and involve the co-researchers in the discussions (see Chapters 5 and 10 of Almekinders *et al.* and Tromp). Emphasising a commitment to transparency, i.e. being explicit about the positions and viewpoints that are taken within the specific research context, is one element in the realisation of such an attitude. Investment in the development of competencies with regard to the cooperation, deliberation and decision-making processes that researchers and co-researchers are involved in, is another.

Reflexivity

Both of these requirements – being explicit and investing in the development of competencies – rely upon a reflexive attitude. Reflexivity entails the capacity to view ourselves as object or from an external viewpoint, like others view us. Up and above that, reflexivity also implies the sensitivity and critical attitude with regard to (most aspects of) our own actions and points of view. We review and asses (the consequences of) these actions and standpoints time and again in the light of new knowledge or information, and reconsider them whenever necessary.

A reflexive attitude is not just about acknowledgement of the fact that researchers cannot keep their own subjectivity out of sight and out of the research process. It is even more about what this acknowledgement means for the development of an adequate (constructivist) methodology. The societal context always influences the scientific practice and the researchers working within this practice. A reflexive attitude implies that this societal impact is made explicit and that a conscious effort is made to bring about social relevant knowledge. The structures and logic that are presupposed by the researcher, e.g. the categories they pose upon society, the relation they presume between attitudes towards certain matters and consequent behaviour, the systematic perspective from which they examine nature and the patterns they find accordingly, also can be studied. As objects of study, they can be changed too. For example, we can adjust our ideas about the relationship between attitudes and human behaviour (which perhaps we assumed to be causal initially but we now have found to be more complicated than that) or about the way systems (don't) work outside the confines of a controlled environment (which may turn out to be in a more unpredictable way than we would have presupposed initially). Research does not have to maintain an internally focused product, i.e. a matter that merely concerns researchers, but can become a matter about which both theoretical experts (the scientific researcher) and practical experts (the stakeholders in the field under study) can judge. This way, research can become a reciprocal process in which theory and practice are interactively connected. To realise this, we have to make a move from one-sided adequacy (i.e. the researcher interprets the object as well as his/her own relation to that object) to reciprocal adequacy in which the stakeholders are involved as a serious party in the research process.

Reflexivity then becomes the explicit acknowledgement of the importance that researchers allow themselves to hear what the co-researchers have to tell them concerning their interpretations of the physical and social reality. This is exactly what the principle of reciprocal adequacy implies. This cannot be brought about when researchers keep forcing their categories upon the researched, but will only evolve when they present these categories as constructions that can be right, but could also be wrong.

Reflexivity and reciprocal adequacy are two closely linked concepts and preconditions for fulfilling the obligation of being explicit about one's position (see Chapters 10 and 11 of Tromp and Beukema). Researchers need to be open and honest about their position within the scientific field, their standpoints both with regard to the subject under study and the research field in general, and about the interpretative schemes that are used to find explanations for the phenomena in the research situation. Furthermore, the obligation to be explicit also calls for a reflexive attitude towards ones own epistemological position (Tromp, in press):

- What presuppositions and assumptions are held with regard to the research object (as can be partly deduced from the interpretative schemes that are used)?
- What knowledge can be relied on and what knowledge is still lacking about the issue under study?
- What kind of justifications are being used to back up the arguments about what is exactly the case and what should be done?
- What arguments can be brought forward to use certain specific research methods and methods for change?
- What are the reasons that some choices are made in the learning cycle of change in general and the research in particular rather than others?

The obligation of being explicit about one's position not only pertains to the professional scientific researcher. Being co-researchers, it applies to the researched party just as well. This can also be viewed as an aspect relating to the criterion of reciprocal adequacy. But just as with the issues of participation and reciprocity, the expectation that the co-researchers will reflect upon their own standpoints and actions cannot be taken for granted. Considering the prevailing ideas about the role of the scientific experts and the presuppositions with respect to the role the research subjects are supposed to take, the requirement that they too take on a reflexive attitude, is not self-evident. It cannot be presupposed but has to take shape during the research itself and is something that needs to be worked on.

When facts and theories on which scientific knowledge is based cannot be viewed as 'given' but must be regarded as at least partly socially constructed, the researcher can no longer be seen as the objective 'know-it-all' who governs and controls the knowledge creation process. The idea that researcher and researched live in two separate worlds, and that the researcher by all means has to prevent him- or herself from being 'contaminated' by the thoughts and actions of the researched, can no longer be held. The knowledge of both subject and object (i.e. co-subject) are indispensable, are complementary in the research process and are decisive for the quality of the results within the learning process that research forms.

Complementary assessment criteria and multi-criteria analysis as we can find, for example, within a jury of the various stakeholders involved in the research process, can function as a method for supporting reflection on the side of the co-researchers. Just as it will enhance reflection on the side of the researchers, for that matter.

The need for reflection is felt both within the scientific domain and within the domains of the policymakers. Reflection on the definitions used in research questions, but also on ethical issues and questions about needs, power relations, and political priorities (see Keune *et al.*, Chapter 2 and Verschoor, Chapter 8 on the debate on genetically manipulated maize in Mexico).

We have to realise though that knowledge can no longer be viewed as an ever-growing source of securities. Instead it must be taken as knowledge that is time and again questioned and further developed by scientists and their co-researchers within specific domains on the grounds of their daily experiences and practices. This exchange between knowledge and practice is context bound and within this exchange there are usually more insecurities than securities. Therefore it is difficult, if not to say impossible, to formulate universal laws or come up with hypotheses on the basis of general statements. This does not exclude the fact that science has something to say on the level of context transcending developments and relations. The aim of generalisation is still upheld (see Tromp, Chapter 10). But general statements about other situations can only be put forward with certain modesty and only with roughly formulated implications, since the starting points for innovation and change are always the more or less reflexive actions of human beings in their own specific context. Every generalisation outside the specific context necessarily will have a tentative character.

All the principles for good cooperation between the researcher and the co-researchers that have been described here can be considered alternative ways to do justice to the criterion of objectivity; they shape the intersubjectivity of knowledge and are an ingredient of the internal validity of the research. Beside the traditional view in which careful observations and logical deductions form the basis for generating objective knowledge we posit a view in which explicit intersubjectivity plays at least an equally crucial role in the internal validation of knowledge. Controlled observations form an important ingredient in making a judgment about what is the case and what should be done to improve the situation. But in the end, the importance of the empirical observations is weighed up and decided upon within the discussions, both of scientists themselves and within the discussions of the scientific researchers and their co-researchers in the field.

Research in action and the actual change of practices

We have seen that vision is the engine for innovation and social change (see Tromp, Chapter 10). But to get the engine running and set a process of change in motion, it needs to be fuelled by the actions of the people involved. We know by now, that merely 'announcing' a change is going to take place, does not make it happen. The reason why innovation or social change are in fact so difficult to realise, is that they can only be brought about when at least a reasonable number of the people involved start to act in a different way. This will not happen 'out of the blue' though. To bring about change, the actual action patterns of the researched should be taken as a starting point. For routines are the background against which the transformation has to be set in motion; we need to know how people act and why they do so, before we can start to make a change (see Chapter 4 of Beukema and Valkenburg, and Chapter 11 of Beukema).

It is not always easy to get a clear picture of how and why people act as they do. In most cases, it is the exception rather than the rule if people are conscious of their own particular actions and the intentions underlying these actions. Their daily actions, including their professional actions, have become so self-evident, that it can be hard for them to explain in words what and why they do it exactly. Moreover, it can be very difficult for actors to grasp the intended and unintended effects of their actions. The first step to be taken in a transformation process is therefore to make individuals reflect on (the consequences of) their actions, and try and let them put into words what they do and why. In sophisticated vocabulary this is called a transformation process from unconsciousness or consciousness on a merely practical level to discursive consciousness, i.e. consciousness on the level that it can be made explicit and be put into words. Again, the interaction between researchers and their co-researchers needs to provide the necessary safety for the co-researchers to dare take up this challenge. Without mutual trust, it is improbable that they will take on the vulnerable attitude needed to reflect upon the consequences of their own actions, which might differ from their intended positive aims.

Even when the preconditions are fulfilled, transformation remains a tricky business. For one thing, each individual will need his or her own time to engage in this process of reflection and gradually change his or her own routines. Some may fall into a defensive attitude, for it is very hard for people to change their habitual actions patterns. Moreover, the individual learning trajectories that will lead to innovation and change will also have to be linked to the learning trajectory that the organisation or community as a whole will go through. If this is not the case, there might be little room for doing the things in another way, and efforts to innovate may suffocate.

This connection between individual and organisational change is where the crucial function of a guiding vision comes to the fore again. The shared mission and the overarching vision of the reorganisation or social change forms the connecting element between the individual learning trajectories and the learning trajectory of the organisation or community. This mission or vision can be inspired by scientific theories, which in turn can be fed by innovating practices (and so on – we mean to refer here to the fruitful exchange between theory and practice). A shared vision is also needed to shed new light on the existing situation, by defining new meanings, power relations and values in a future-oriented, but nevertheless realistic way. In this sense a shared vision is a precondition for quality and validity, or even stronger: sometimes a condition sine qua non (see Van der Haar's chapter). In the same way, a research project can enhance the process of change by bringing together the different kinds of knowledge and learning trajectories. It can fulfil the function of relating individual learning trajectories to organisational learning trajectories (see Figure 1 in Chapter 4 of Beukema and Valkenburg, where knowledge development forms the connecting pillar between both types of learning trajectories). Moreover, by putting (extra) demands on the change process, a facilitating and supportive (scientific) research project can also enhance the quality and validity of the results of the overall learning process.

Interaction between actors on various levels

Research in communities or organisations is always research in existing fields of relations, where different actors are involved in processes of sense making and (re)producing power relations. The researcher not only has to be aware of these relations, but also of the fact that (s)he becomes part of them and has to deal with them. The position of the researcher often is multi-layered, i.e. the researcher needs connections with actors from different parts of the organisation or community. Some room to manoeuvre within the organisation or community is therefore essential: the management or the citizens have to give the researchers enough space to disagree and carry their own responsibility in the change process (see Van der Haar, Chapter 7).

In forms of research within organisations in which the learning of co-researchers is an objective, trust between management and researchers is a precondition in order to build up relationships of trust with other members of the organisation. The researchers also have the responsibility to build such a relationship of trust with the operational management and the professionals responsible for the primary process, i.e. the 'workers on the floor' (e.g. community workers, street level workers, those who provide care or other services). This relationship of trust cannot be taken for granted, since there are differences in power between the researchers and the co-researchers. Therefore, the aim is not equality but equivalence: both parties have their own position in the process of action, learning and building knowledge and both should value and respect each other in their particular positions (see Chapters 6 and 4 of Roose and De Bie, and Beukema and Valkenburg).

Research within communities can vary greatly in scope: the community can be a group of employees in an organisation, a local neighbourhood, rural area or municipality but also the world community. Whatever the scope may be, the researcher or research team has to take into consideration the different levels on which the learning processes within the research are taking place. To manage such a complex of processes successfully, a reflexive input of flexibility is needed, i.e. the question of when the input of which resources is in place, needs to be carefully considered. This calls for consideration of the input:

- in time: the planning of the learning processes needs to be adjusted to the needs and motives of the individuals and/or the particular organisations or communities involved;
- in space: cooperation needs to be sought and networks need to be established between those individuals, organisations and/or communities that play a important role with regard to the theme that forms the central research subject;
- in function: differentiation and diversity should be allowed, for only then can a transformation process gain power and strength;
- in thought: researchers and co-researchers need to develop an open attitude towards innovative points of view that can inspire and feed the change process;

How can we bring about innovation and social change through research?

- By setting up research as a (mutual) learning process
  - which is based on trust between the researchers and the co-researchers
  - in which an acceptable level of participation of the co-researchers is enhanced
  - in which reciprocal adequacy of the research findings forms a guiding principle
  - and in which the attitude of both researchers and researched is characterised by reflexivity
- By creating a learning environment in which
  - personal and collective learning trajectories are adjusted to each other
    - ▹ starting with the daily actions and routines as a basis for change
    - ▹ aiming to get from unconsciousness or consciousness on a merely practical level to consciousness on the level that the intentions and (un)intended consequences of individual's actions can be made explicit and be put into words
    - ▹ within a research project in which the different kinds of knowledge and learning trajectories are brought together
    - ▹ and in which the shared mission and the overarching vision of the reorganisation or social change forms the connecting element between the individual learning trajectories and the learning trajectory of the organisation or community
  - local, national and global perspectives are taken into consideration
    - ▹ research takes place on different levels, within organisations and outside
    - ▹ there is interaction between actors on those different levels in process
    - ▹ a reflexive input of flexibility is realised in time, in space, in function and in thought.

- in action: everybody needs to act according to their own strengths and capacities, but always act on what they say (practice what you preach);

12.4 Why do we need this new kind of research and what do we need for it?

Professionalism redefined; implications for the research competencies

The contributions in this book show us telling examples of the challenges our contemporary society is currently facing: issues on the intersection of various disciplines, characterised by complexity in multi-stakeholder settings. Addressing these issues requires researchers who are able to communicate with all these stakeholders while generating *and* applying the (new) knowledge that is gained by continuously relating theory to practice. This means, that they need to be able to engage themselves with the practices under study, and get a grip on all the complex interactions and 'power games' that are going on there. At the same time they have to take on the necessary objective attitude to transcend the particular situation and relate the learning process that is going on in the concrete context to a more general level (see Chapter 4 of Beukema and Valkenburg).

The participation of stakeholders in the research calls for academics with a broad range of competencies. Beside the research skills that are traditionally emphasised, a researcher nowadays is required to be able to learn within a dynamic process, integrate the various kinds of knowledge that are being generated on various levels and are being assessed by a plurality of evaluative frameworks, and to transfer and apply this knowledge in a way that fits the actual situation (see Chapters 5, 8, 10 of Almekinders *et al.*, Verschoor and Tromp). They are increasingly expected to be effective team players in interdisciplinary research, and able to design and co-facilitate participatory processes aimed at learning, development and change in organisations or communities. Within these processes, they need to be able to relate empirical knowledge about 'matters of fact' to normative claims about which values are worth striving for, how these can be accomplished, what means can or should be used to realise these aims and how all this can be brought together in strategic action plans. Meanwhile, they have to be sensitive enough to handle precarious power relations and compare the different ways of justification and legitimisation that are being used by different stakeholder groups. All this in a context where flexibility is a condition without which the whole research process would collapse in no time.

However, up till now, academic curricula have not paid much attention to providing training in the competencies that relate to successful human interaction. Researchers need to be equipped with the social skills to effectively use methods and techniques that are related to these aspects of human interaction and all that it entails. Partly to acquire the competencies to apply the methods and techniques to enhance building team relations and to facilitate learning processes. But also because they need the basic soft skills related to being a good communicator, a team player, etc.

For academic institutions to pay a contribution to the actual and future challenges of the world, their curriculum should focus on producing this new type of professional. What we need are researchers with the competencies to build bridges between science and society. Scientists that realise that they cannot claim to possess the exclusive though all-inclusive knowledge to find *the* solutions for the complex questions our societies are facing. Academically trained professionals who know how to translate scientific knowledge, with all its subtle nuances like levels of significance and probability and with all the uncertainties and limitations connected to it, to the worlds of the politicians, policymakers and managers in which issues are framed in a rather different manner. Experts who can help bridge the gaps between the diversity of actors involved in a research process, by introducing clear procedures for observation, cooperation, discussion, deliberation, assessment, validation, decision making and action planning (see Keune *et al.*, Chapter 2).

As Banathy remarked back in 1999: 'Today we live in an era of massive societal changes and transformations that are reflected in the new realities of the post-industrial information knowledge age. These changes touch the lives of every person, family, community and nation and affect the future of humanity. However, we are entering the 21st century with educational systems that were designed during the 19th century. Improving or restructuring

Why do we need this new kind of research and what do we need for it?

- Because are contemporary society is faced with challenges
  - that cannot be studied from a monodisciplinary perspective but must be viewed in all their complexity and approached from an interdisciplinary perspective
  - that involve various groups of stakeholders, each with their own interests in the issue at hand.
- Therefore, our society is in need of a new kind of expert professional, who is able
  - to combine and integrate the various disciplines involved in the challenges at hand
  - to deal with the complex interaction with co-researchers, often consisting of a variety of stakeholders
  - to handle the tricky balance they need to find between engagement and objectivity
  - to relate to both systematic academic knowledge and 'messy' lay knowledge
  - to integrate empirical and normative knowledge, combining various regimes of justification and legitimisation used by different groups of stakeholders
  - to facilitate the learning process in such a way that new knowledge can be put into action, thereby
  - enhancing the implementation of scientific knowledge into societal practices
  - to bridge the still existing gap between the academy and the world of policymakers, managers, clients and citizens
  - to help formulate a constructive perspective on highly complex issues like those mentioned in the contributions, and other challenges facing our world community.

these systems, which were created based on the design of the industrial machine age, do not and will not work in the post-industrial knowledge era, only a radical and fundamental change of educational perspective and purposes and the reconceptualisation and redesign of our educational institutions will satisfy the merged new realities.' (Banathy, 1999: 144)

References

Banathy, B., 1999. Systems thinking in higher education: Learning comes to focus systemic. Research and Behavioral Science 16: 133-145.

Gibbons, M., C. Limoges, H. Nowotny, S. Schwarzman, P. Scott and M. Trow, 1994. The new production of knowledge: The dynamics of science and research in contemporary societies. Sage, London.

Kunneman, H., 2005. Voorbij het dikke-ik. Bouwstenen voor een kritisch humanisme. SWP, Amsterdam.

Tromp, C., 2008. Criteria for a reflexive approach to knowledge production future. In: J. Zeelen, B. Boog, M. Slagter and J. Preece (eds.), Towards quality improvement of action research: Developing ethics and standards (in press).

Glossary

Words in a description which are printed in italics, are also explained separately.

Abduction A recurrent process in which both *induction* and *deduction* are used to formulate hypotheses about reality.

Action learning A voluntary, participant-oriented development process aimed at finding solutions for real, systemic organisational work- and learning problems in labour situations, meanwhile trying to do justice to the principle of democratic values and team learning in an environment characterised by trust and authenticity. Corner stones of action learning are:

1. the problem (that needs to be reformulated into a project);
2. the client (i.e. the owner of the problem);
3. the other persons involved (i.e. the participants in the project team);
4. the advisor (i.e. the action-learning expert);
5. the coaches (i.e. specialists with regard to the subject at hand);
6. the meetings and workshops.

Action research Form of scientific knowledge-gathering which seeks to bring together action and reflection, theory and practice. The knowledge production process is viewed as a participatory democratic process concerned with developing practical knowledge in the pursuit of worthwhile human purposes. Within this approach the researchers and clients collaborate in the diagnoses of a problem and in the development of practical solutions based on the diagnosis (Coenen, 1987, 1998; Reason and Bradbury, 2001: 1; Bryman, 2004: 537).

Activist research A research practice that starts from an alignment with an oppressed, disenfranchised, or marginalised collective subject, whose members are engaged in struggle for relief from oppression, for rights, in general for improvement of their situation. In many cases, researchers work directly on or with the people with whom they are aligned, accompanying them in their struggles and developing a research agenda and action strategies from this experience. Activist research draws on theoretical work directed toward an understanding of how relations of knowledge/power are (re) produced in society, but also how they can be challenged. These theories include amongst others, feminist theory, Marxism, Foucauldian analysis and critical race theory. Furthermore, it is informed by a diverse body of work that moves from critique to the ability to inform strategic thinking and direct action on a given issue or problem (cf. Austin Concept Statement on Activist Anthropology, October 2003).

Actor causality	The autonomous position of human beings with regard to most (though not all) factors in its environmental surroundings. Every subject is an autonomous causal centre of his/her own actions, in the sense that the question of the influence of the factors in his/her environment is in the end determined by the subject (Coenen, 1987: 143).
Actor-network theory	An approach (originally to sociotechnical analysis, but increasingly used in other fields) that sees all entities (whether human or non-human, discursive or non-discursive) as relational effects. Its main object is the tracing and re-tracing of these relations, without giving analytical priority to well-know ontological categories such as 'nature', 'society', 'discourse', 'agency' or 'structure'.
Biomarker	A substance used as an indicator of a biologic state or of substances in the body. It is a characteristic that is measured and evaluated as an indicator of normal biologic processes, pathogenic processes, or pharmacologic responses to a therapeutic intervention. A biomarker of exposure measures exposure to certain substances, e.g. pollutants; a biomarker of effect measures health effects, e.g. of environmental pollution.
Biomonitoring	See human biomonitoring.
Boundary critique, boundary work	Reflection on the borders of what falls within and what is excluded from the environment that is studied (which is often viewed as a system). The underlying values or criteria for the distinctions and consequently the selections that are made can be contested, and often are, not only by scientists themselves but also by experts outside the scientific field (cf. Midgley 2000; Gieryn, 1995: 405-406).
Classes	In Marxist analysis class refers to the relationships of people to the means of production, and this entails exploitation and conflict.
Co-operative inquiry	Form of research in which two or more people study a topic through their own experience of it, using a series of cycles in which they move between this experience and reflecting together, in a reciprocal relation, on their knowledge and experience. Each person is co-subject in the experiential phases and co-researcher in the reflection phases (Heron, 1996).
Collaborative research	A deliberate set of interactions and processes designed specifically to bring together those who study societal problems and issues (researchers) with those who act on or within those societal problems and issues (decision makers, practitioners, citizens). Collaborative research implies the involvement of non-researchers in the conduct of research but excludes partnership when it is just for funding or access to research sites. In this context it is not necessarily, although it is often practically, multidisciplinary collaboration (Denis and Lomas, 2003).
Common sense	Everyday knowledge that is taken for granted including interpretative schemes and expectation patterns of social actors about both physical and social causal relations in the world.

Communicative symmetry	Social conditions related to an ideal argumentative situation (ideal discourse), implying an equal distribution of chances to take part in the discussions, to take the stance and either defend or criticise particular statements.
Complexity	Complex systems have certain important characteristics, including being constituted by a large number of elements interacting richly, locally and non-linearly, containing feedback loops and being far from equilibrium (Cilliers, 1998).
Co-production of knowledge	See collaborative research and co-operative inquiry.
Credibility	The arguments and the processes necessary for having someone trust research results (in pragmatic action research formulated as alternative for validity in conventional social research). Two different types of credible knowledge are distinguished: internal credibility to the group generating the knowledge. External credibility is knowledge capable of convincing someone who did not participate in the inquiry (Greenwood and Levin, 1998: 80-81).
Critical theory	In social science, critical theory has two meanings. The first is that theories about society should be oriented towards a critique (and change) of positivistic science and society. This is a stance best elaborated by the Frankfurt School and (later) Bourdieu, Althusser and Foucault. Another version of critical theory derives from Kant's *Critique of Pure Reason* and Marx's critique of political economy, and can best be characterised by Kant's project of examining and establishing the limits of the validity of knowledge – an endeavour that was further developed by Marx' critique of ideology which he linked to the practice of social revolution. Both versions of critical theory have in common an emancipatory interest and the critique of domination.
Crystallisation phase	Phase within *(exemplary) action research* in which researchers and co-researchers try to rank the problems identified by further articulation of the research question. Ideally, this should lead to the selection of one central problem theme (an *exemplar*) that, if successfully tackled, should lead to improvements within the problematic situation. To be able to realise this aim, an action strategy and working plan with regard to the central problem theme is formulated (Coenen, 1987).
Deduction, deductive	Inferring specific hypotheses from general theories or laws. An approach to the relationship between theory and research in which the latter is conducted with reference to hypotheses and ideas inferred from the former (cf. Bryman, 2004: 538; Walliman, 2005: 10-11, 193, 222, 432).

Delphi method	Delphi involves an iterative survey of experts. Each participant completes a questionnaire and is then given feedback on the whole set of responses. With this information in hand, the respondent fills in the questionnaire again, this time providing explanations for any views they hold that were significantly divergent from the viewpoints of the others participants. The explanations serve as useful intelligence for others. In addition, the respondent may change his/her opinion, based upon his/her evaluation of new information provided by other participants. This process is repeated as many times as is useful. The idea is that the entire group can weigh dissenting views that are based on privileged or rare information. Thus, in most Delphi processes the mount of consensus increases from round to round (Slocum, 2003).
Demand-driven approach	The transformation of an institutionally driven, supply-oriented approach (of care or other policy fields) towards an approach that takes the individual client as a starting point and director of the process. A demand-driven approach primarily means that policies link up with and do justice to the reflexive projects of individual people in a reciprocal way (Beukema and Valkenburg, 2007).
Democratisation	Process in which societal decision making and the institutionalised practices based upon these decisions are taking place – as much as possible at least - in open, publicly accessible discussions in which people can let their voices be heard and can try to exert influence to get subjects they find important on the social and political agenda.
Deroutinisation	Reflecting upon the daily interactions that are usually just taken for granted and taking distance from standard habits and action patterns to try and prevent the (often unconscious) reproduction of existing societal structures (Giddens 1979: 220; Coenen 1987: 150, 167).
Determinism	A stream of thought suggesting that the actions of human beings are (almost) exclusively determined by external factors, for instance societal relations or structures.
Dichotomy	Placing two concepts opposite of each other as if they are essentially contrary notions.
Discursive consciousness	Being able to put the knowledge and insights that we have about our environment, about our being in the world and about how our actions relate to bigger relations, structures or systems, into words.

Discourse	Some use discourse to refer to an ideal discussion, in which participants all have an equal chance to bring forward validity claims with regard to the truth, justice and/or authenticity of their statements and expressions, and all have the opportunity to fundamentally criticise the validity claims of others or the presuppositions underlying these statements (Habermas, 1973: 253, 1981). For others, discourse refers to the dominant language game that determines what people can say and cannot say in particular context at specific times (Foucault, 1982; Derrida, 1983; Lyotard, 1987)
Double hermeneutics	The fact that scientists in the study of social phenomena not only have to rely on interpretation to be able to gain an understanding of the actions of the researched or co-researchers, but in their efforts to give meaning to the action under study also have to rely on the same interpretative frameworks as those actors themselves do (Giddens, 1976: 155).
Dualism	Framing in an opposition, i.e. placing concepts against each other as direct opposites in a dichotomy, as if they are pole and antipole (e.g. subject - object; action - *structures*; *determinism* - *voluntarism*).
Emancipation	The liberation of people – by themselves and others – from unequal relations in which access to important resources is inhibited or denied, or in which the results of their actions are to a large extent determined by others. Strengthening people's societal position by increasing their competencies, (self)-consciousness and insight into (the intended and unintended consequences of) their own actions (*empowerment*) and by linking a broader range of action scripts and strategies to these acquired insights.
Empiricism	*Empiricism* is '... a family of traditions in the philosophy of science which argue that scientific truths grow out of, and are properly generalised from, appropriate empirical observations' (Law, 2004: 16). See also *logical empiricism* and *positivism, positivistic method*.
Empowerment	Giving individuals more influence and power to direct their own lives by augmenting their insight into their options for action and the way in which their actions help to reproduce existing action patterns, and also by strengthening their competencies and teaching them the practical skills to bring about change and improve their situation.
Enlightenment	In its simplest sense the Enlightenment was the creation of a new framework of ideas that emerged during the 18th century about man, society and nature, and which challenged existing conceptions rooted in a traditional world-view, dominated by Christianity. It is probably futile to pin down a single definitive group of people or set of ideas which can serve as *the* Enlightenment. Nevertheless, famous *philosophes* (the initiators of the Enlightenment) such as d'Alembert, Diderot, Voltaire, Montesquieu, Hume, Locke or Rousseau would have agreed that reason, empiricism, science, universalism, progress, individualism, toleration, freedom and secularism were its main ingredients (cf. Hamilton, 1992: 21-23).

Epistemology Theory of knowledge; theory on how we people can know reality, how we can gain adequate knowledge and what is a good legitimating strategy for our opinions (Harding 1987: 2).

Ethno-methodology An approach which studies the methods that people use to arrive at an understanding of each others practice, and the way they organise their action around that understanding. In ethnomethodology the individual aspect is hardly thematised; the relation between the interpretation of the actor on the one hand and his/her behaviour on the other hand, is not an object of study.

Event causality causal influences related to conditions and circumstance of which actors have no knowledge, of which they themselves are not conscious, while those factors can have an important influence on their actions (Giddens, 1976: 154-159; Coenen, 1987: 195-196).

Exemplar The most important theme, according to researchers and co-researchers, that should be tackled first in the research. For a theme to become an exemplar, there are three criteria: (1) It should be a central theme within the organisation; to handle it is a condition for handling other themes. (2) It should be realistic, in the sense that it must not only deal with problems, but also with strengths and realistic aims in the given context. (3) It must be relevant for the learning process of the organisation; what is learned in one situation must have exemplary value for others (Coenen, 1987, 1996).

Exemplaric phase Phase within *(exemplaric) action research*, in which the execution of the action plan that is linked to the *exemplar* takes places and the effects of these actions are evaluated (Coenen, 1987, 1996).

Exemplary action research Research model in which researchers and co-researchers, on the basis of continuous data collection, discussion and joint action, try to select an *exemplar*. Presupposition is that tackling this exemplar, i.e. the most central and essential problem theme, can lead to a solution to the problematic situation under study and that this can have an exemplary –function for other, comparable research situations (see Coenen, 1987, 1996, 1998).

Family resemblances The characteristic, visions or paradigms that are shared by different schools of thought.

Facilitating Supporting learning processes, for instance by enhancing the experience with how changes in existing action patterns and daily routines can lead to improved practices. By offering room and the resources to go through personal learning processes, individuals can develop a broader repertoire of action strategies. Moreover, by clarifying the relation between the individual and collective actions and making people conscious of this relation, the motivation for changed and changing actions can be enhanced, and more general practices, processes or learning processes can be developed.

Farmer field schools A group-based learning approach in which farmers learn about their agricultural production and the interaction with its environment. The approach was originally used as a way for farmers in South East Asia to learn about pest management in rice. Lately many more topics are addressed in a Farmer Field School(FFS) model, by many organisations (see also Wikipedia)

Feminism Perspective within social theory that studies gender in general and gender differences and similarities and gender relations in particular, as well as the influence hereof on the societal relations.

Feminist empiricism See also *Standpoint feminism*

Forum Group of experts that judge the scientific validity of the generated knowledge.

Generalisation, generalisability The degree to which findings in one particular research context can also be taken to have explanatory power for another, comparable situation (cf. Bryman, 2004: 76-77, 287; Walliman, 2005: 13, 433).

Grand narratives Generally understood as a set of related, hegemonic discourses strongly embedded in modernist thinking. In social science, the most influential discourses of this kind are strongly related to images of unity, purity, and order. Unity refers to the conception of 'the social' as a centred whole, and the Grand Narratives associated with unity focus on what unified this totality of 'the social' (which was traditionally taken to be the same as 'society' in the sense of a nation-state with clear-cut boundaries). For Marx, the big unifier was capital, for Durkheim it were shared norms and values, while for Weber it was rationality. Grand Narratives associated with purity, in turn, are related to the idea of 'the social' as something homogeneous (or 'pure') in the sense that it can be considered sui generis (cf. Durkheim's [1938: 110] formulation: 'the determining cause of a social fact should be thought among social facts.'). Finally, Grand Narratives arising from a preoccupation with order generally seek to detect the 'laws' of social change (which for Durkheim were linked to organic solidarity, for Marx to the logic of the economy, while for Weber they were tied to the iron cage of bureaucratised society (cf. Albertsen and Diken, 2003: 2-3).

Human bio-monitoring Monitoring activities in human beings, using biomarkers, that focus on environmental exposures, diseases and/or disorders and genetic susceptibility, and their potential relationships.

Debates about highly confused issues (e.g. mad cow disease, the hole in the ozone layer) in which both experts and lay persons participate and in which facts and values become entangled to such an extent that it is no longer possible to distinguish between two successive stages: first, the production and dissemination of information and knowledge, and second, the decision-making process itself (Callon, 1998: 260).

Induction, inductive	A logical and methodological way of reasoning in which from (the integration of) particular, separate empirical observations more general theoretical statements are deduced (e.g. deducing from a large amount of separate observations of white swans that all swans are white) (cf. Bryman, 2004: 8-11, 540; Walliman, 2005: 11, 433).
Innovation	Process in which technological inventions (e.g. new kinds of artificial fertilisers, chips, fuels) or social renewal (e.g. the globalised economy, or the fact that people now continuously re-invent their own self-images) can lead to a shift or even radical change in existing production processes and social practices
Interdisciplinary approach	Form of research practice in which scientists from different disciplines work together and several perspectives are combined and integrated by viewing them in their mutual coherence (or incoherence, if there is a problem!). The synthesis of the different perspectives takes place during the research, in the process of formulating a problem definition, methodology and interpretation of the results that are acceptable to all the researchers of the various disciplines. Interdisciplinarity is most often a temporary stage which either dissolves or leads to the founding of another new discipline (Weingart, 2000).
Internal reflective questionnaire	Questionnaire put individually to participants in a process of dialogue and/or cooperation, reflecting on the issues at stake, for example with regard to content of a problem or process aspects. The idea is that the collection of views and ideas on an individual level enriches the discussions with diversity from different perspectives. In group discussions this diversity may 'disappear' unnoticed. This diminishes opportunities for the actors involved to learn from other perspectives that (in that case) partly remain in the blind.
Intervention research	A term for research that comprises research on interventions as well as research through intervention
Intradisciplinary science	See Interdisciplinary approach
Local Agricultural Research Committees	Local Agricultural Research Committees (known by their Spanish acronym, CIALs) are locally organised groups of farmers who voluntarily carry out agricultural experimentation, to test and develop technology. The results of their work are serving the rest of the committee.
Logical empirism	See *positivism, positivistic method.*
Method	The rules and techniques for the collection and analyses of theory relevant facts, for instance making observations and scoring them within an experiment or developing questionnaires and interpreting the answers within a survey.

Methodology	A theory or analysis about how research should be done; system of logical or philosophical principles for structuring theory relevant facts (Harding, 1987: 2).
Modern	Label for the (form of) science that legitimises itself by referring to meta-discourse, i.e. to a discourse that legitimises its own status or, put differently, to the discourse that is called 'philosophy' (Lyotard in Rorty, 1985: 161).
Modernity	According to Latour (e.g. 1993). the word 'modern' designates three sets of entirely different practices which must remain distinct if they are to remain effective, but have recently begun to be confused. The first set of practices, by 'translation', creates mixtures between entirely new types of beings, hybrids of nature and culture. The second, by 'purification', creates two entirely distinct ontological zones: that of human beings on the one hand; that of nonhumans on the other. Without the first set, the practices of purification would be fruitless or pointless. Without the second, the work of translation would be slowed down, limited, or even ruled out. Finally, the third set of practices (which can be held responsible for justifying relativism, domination, imperialism, false consciousness, syncretism) consists of making a distinction between 'Us' moderns (who separate between Nature and Culture) and 'Them' – the pre-moderns, or 'backward' cultures who do not make this separation.
Modernisation	The mutually related developmental processes of: 1. increasing differentiation, not only in the domain of the economy (division of labour), but in all domains of society (beside the formation of capital and an increase in labour productivity one could think of the installation of a centralised political power and, consequently, the constitution of national states; the introduction of the right to political participation, urban life forms and formal education); 2. secularisation, i.e. the fact that on the cultural level traditional norms and life patterns are no longer taken for granted (Beck, 1986: 206; Habermas, 1984: 10).
Mode 1 knowledge	Cognitive-instrumental knowledge, i.e. knowledge to find theoretical explanations and practical solutions for scientific problems, usually monodisciplinary in focus.
Mode 2 knowledge	Knowledge produced in a process that is context-driven, problem-focused, and transdisciplinary (Gibbons *et al.*, 1994).
Mode 3 knowledge	Knowledge concerning personal learning processes in which moral and existential issues are also addressed.

Multi-criteria analysis A method that mathematically compares several options (for example, products or policy measures) on different criteria, thereby structuring a complex process of decision making based on relevant information and possibly different actor-perspectives (e.g. a group discussion). The options are compared based on a combination of the values on each criterion and the relative weight criteria have for actors involved in the process (e.g. the relative importance of economic versus environmental or social aspects).

Multidisciplinary approach Form of research in which several perspectives are put beside each other. The synthesis of the different perspectives takes place post hoc, when the individual results of the disciplinary research are combined to find an answer to the problem.

NGO A Non Governmental Organisation (NGO) is a non-profit group or association organised outside of institutionalised political structures to realise particular social objectives (such as environmental protection) or serve particular constituencies (such as indigenous peoples). NGO activities range from research, information distribution, training, local organisation, and community service to legal advocacy, lobbying for legislative change, and civil disobedience. NGOs range in size from small groups within a particular community to huge membership groups with a national or international scope.

Ontology Theory on matter and beings, i.e. on the constituents that form reality and structure it.

Orthodox consensus Rather broadly shared agreement within science, at least until the end of the sixties but still very influential today, that the social scientific enterprise can best be modelled after the natural scientific way of research, and that criteria like objectivity and generalisability should be the guiding principles.

Paradigm A system of presuppositions that is taken for granted within a research approach or research school and that forms the framework within which problem solutions are sought (Kuhn, 1962, 1970).

Participatory extension approach A participatory extension approach (PEA) was developed in Zimbabwe. It integrates elements of participatory technology development (PTD) and approaches such as action learning. It involves a transformation in the way extension agents interact with farmers (Hagmann *et al.*, 1999).

Participatory action research (PAR) Approach of Action Research developed in the 1970s in Third World countries to investigate reality in order to transform it through praxis (Borda, 2006). Also called: Southern PAR (Greenwood and Levin, 1996: 175).

Practice cycle A cyclical scheme with different procedural steps, actors and roles for (different phases of) work (deliberation, research) in progress, e.g. a decision-making process.

Plurality, Pluriformity The position that within science several perspectives and approaches can exist beside each other. Also: The position that social and physical phenomena can be made up from different materials or substances.

Positivism, positivist method Scientific approach that came into being at the end of the 1920s, representing the stance that the best way for both natural and social scientific research is to *inductively* deduce hypotheses via systematic observation and subsequently enforce these hypotheses via verification, i.e. via the positive function of argumentation (cf. Popper, 1963: 229). In this vision, the scientific system of 'true', i.e. logical correct and consistent statements could only be deduced from the 'positum' (that which is proposed), i.e. from the empirical observable facts and mathematical or logical relations (cf. Walliman, 2005: 203).

Postmodern, postmodernism A way of thinking which is understood as a break from the modern way in which the World understands itself, in how it understands cultural formations like science and art, knowledge and power. It breaks with the modern thinking of control, in which a vision on the world is characterised by an orientation on the position of the subject (as source of control), a search for unity (as theoretical form of control), hierarchy and a natural way of conformation to norms and essences. In contrast to this there is a plea for deconstruction, heterogeneity and pluriformity.

Postmodern feminism An approach to feminist theory that is sceptical about universal Grand Narratives – for example about the 'nature' of women. It posits that all knowledges are specific and produced locally through struggle and made apparent through language. Because of their specificity, there is no such thing as an 'essential' woman: 'sex' and 'gender' identities are always fractured along lines of sexual preference, ethnicity, or class position. In terms of its agenda, post-modern feminism seeks forms of knowledge or 'truths' which builds links between fragmented identities, interferes with existing injustices, and engages in political contestation to further change (cf. Butler, 1999; Kristeva, 1991; Irigaray, 1985; Haraway, 1989).

Post-structuralism The idea that there is not one single deep structure or 'grammar' that governs action or language but that rather in every era there exist a multitude of 'grammars', which means that one can identify these 'discourses' as well as the way in which they are created and deployed. The argument then is that everything 'social' (people, social interactions, texts, architectures) can be seen as a relational effect, ordered in terms of a discursive syntax.

Postulates Basic presuppositions, constitutive principles defining a specific praxis or approach of reality that do not allow any deviation without being excluded from that praxis or approach of reality.

Pragmatism Scientific position stating that the meaning of concepts and the truth of statements can only be captured in terms of the social practices of the linguistic communities in which they are used.

Pragmatic action research	Knowledge construction processes that involve both researchers and local stakeholders in the same learning action process, thereby fulfilling both a participative democratic ideal and achieving knowledge generation through learning from action. An approach of action research that is grounded in the pragmatic philosophical tradition (Levin and Greenwood, 2001).
Practical consciousness	The actual everyday knowledge of people that has come to be taken for granted and of which they are hardly conscious of having, let alone able to put into words.
Rationality	Knowledge generated by humans for reasoning. According to the traditional point of view this knowledge is bound to the senses and our cognitive capacities, as well as to demands of purity (for instance formal logic), exactness and accuracy. This way, the hope was that we could find general laws that would transcend time and context. In a more modern, broader defined concept our informal reasoning capacities and fruitful experiences from practice are also taken into account. In this view, knowing is not limited to 'facts' but also involves norms and values and need interpretations. This way, we can come to knowledge that is bound to time and context, but which nevertheless has explanative power for and can be useful in comparable situations.
Reciprocal adequacy	Principle stating that the adequacy of knowledge is enhanced by feeding the interpretations, descriptions and explanations that researchers formulate on the basis of the interpretations, descriptions and explanations of the researched, back to the researched and put them before them to check (Coenen, 1987: 210 ff.). The principle of reciprocal adequacy can be viewed as the equivalent of the criterion of (internal) *validity* as used within the orthodox consensus.

Regime of justification A model developed by French sociologists Luc Boltanski and Laurent Thévenot to understand political disputes. By studying the details of how ordinary people engaged in disputes about right and wrong justify their actions, these authors have been able to identify six different 'regimes of justification' (which they call Cités in French). The novelty of their approach is to have proven that each of those regimes is complete, although utterly contradictory with the others. In other words, it is possible to demonstrate that in contemporary [French] society people engaged in disputes may ascend to six different overarching principles, each of them engaging a full-fledged and coherent definition of what humanity should be. Each regime is the result of a long history of political philosophy, and has now become an everyday competence easily activated by every member of society. Each of them defines through trials a scale of right and wrong, that allows one to pass judgment and to settle disputes. Each of them – and this is the great strength of the model – has the capacity to denounce the others because they lack morality or virtue (Latour, 1998: 224).

Reflexive modernisation Process in which the principles that currently form the guiding lines of the societal developmental processes are no longer taken for granted but are put to the test and are analysed on possible inherent contradictions, so that we get a clear perspective on their meaning for the actual societal situation, and are in a position to adjust our guiding principles where needed (Beck, 1986: 253-254).

Reflexivity The capacity to view our self as an object or from an external point of view, as others would see us (Côté and Levine, 2002: 223). The susceptibility with regard to (most aspects of) our own actions and the standpoints that we take, which are reconsidered time and again and are readjusted in view of newly gained knowledge or information (Giddens, 1991: 20).

Relativism In general: the view that the way in which existence and perspectives, or the validity of knowledge and claims, and the correctness of norms and values are formulated, totally or partly depend on the view of an individual, a group a community, society, culture a historic period, ecological conditions, a linguistic system, a paradigm, a conceptual system, a world vision, or another context. In the narrower sense: the view that no general valid or universally accepted c.q. objective criteria for knowledge, truth, values, rationality, etc. exist, which implies that different cultures or different scientific approaches cannot objectively be compared, since there are no independent measures. (Smaling, 1987: 174-175).

Relevant actors Actors that may have a relevant viewpoint, relevant knowledge or a stake with regard to an issue in a process of knowledge production, assessment or decision making. The relevance is defined by the actors actively involved in or responsible for designing, steering or taking decisions in the process.

Risk communication The communication about risks, such as a natural hazard or environmental risks for public health. Traditionally this was conceptualised as one-way communication between experts from science, industry or governmental institutions towards the public. Modern risk communication emphasises the need for two-way communication in which the voice of people not considered as professional experts are also taken seriously as is the need to involve aspects of risk perception.

Risk perception The subjective judgment that people make about the characteristics and severity of a risk, for example a natural hazard or environmental risks for public health. Public health risks related to environmental pollution are scientifically very complex. These risks are also socially complex: they are interwoven with our way of life, with our norms and values. Different perceptions of risk are related to many factors. Next to scientific factors, also social factors are very influential.

Scientism *Scientism* has different meanings. Here we use the one given by Webster's Dictionary: '1: methods and attitudes typical of or attributed to the natural scientist 2: an exaggerated trust in the efficacy of the methods of natural science applied to all areas of investigation (as in philosophy, the social sciences, and the humanities.)' The corollary of scientism is the belief that scientific argument should always be weighted more heavily than other forms of wisdom, particularly those which are not (yet) justified from within science (i.e. lay knowledge).

Semiotics A linguistic theory that says that the terms in a language are significant not because they relate to the objects which they claim to describe, but rather because of their relations with one another, like in 'mother/father', 'nature/culture', 'black/white', and so on (think, perhaps, of a *network* of terms related to one another in some kind of grammar).

Social complexity The complexity of social systems, aspects and interactions or an approach to social phenomena that tries to analyse a social system as a complex system.

Social Structure Generative rules and resources, which provide the formulas, procedures and means for action and which are organised as the relatively constant and stable elements or properties of *social systems*.

Social system Interaction forms that show a similar structural pattern, or relations between (groups of) actors that can be analysed as a recurrent social practice. A social system has got structures, or put differently: structural and structuring properties, though it is not a *structure* in itself (Giddens, 1979: 66).

Standpoint feminism In analogy to Marxism, standpoint feminism argues that society is ruled by men who – by way of a patriarchal ideology - conceal exploitation and distort reality. Truth is only visible to exploited groups (such as women), and hence undistorted, non-ideological and thus valid science can only be practiced by members of such exploited groups. (cf. Benhabib and Cornell, 1987).

Stakeholder A person or organisation who has an interest in, is involved in, affects or can be affected by, an issue (e.g. an environmental problem), a product, a project, or by actions of an organisation (e.g. a company or governmental authorities).

Structuration Process by which *structures* are constituted by the actions of people and vice versa: by which people's actions are structurally constituted (Giddens, 1976, 1979, 1981, 1984).

Thematic phase Inventory phase within *(exemplary) action research*; starting from the central research question the researchers and co-researcher search for the most important themes within the research field with regard to the problem at hand (Coenen, 1987).

Trans-disciplinarity Characterisation of a type of knowledge that includes interdisciplinarity but goes a step further and transcends the margins of science, understood as the organised production of knowledge in university-based disciplines. More precisely, transdisciplinarity transcends disciplinary science in two ways. First, it transcends science in relation to the persons involved, i.e. its also involves non-scientists in the production and/or evaluation of knowledge. Second, it transcends science in relation to the problems involved, i.e. it explicitly orients its knowledge production not only around disciplinary problem definitions but also around other definitions, derived from pressures, 'applications' or from societal stakeholders, no matter whether these be government agencies, private firms, NGOs or social movements (Gibbons and Nowotny, 2001).

Triangulation The principle of approaching a research question from different angles (theoretical triangulation), with various techniques (methodical triangulation), possibly using various sorts of findings (data triangulation) by different persons (researcher triangulation) (Bryman, 2004: 275, 545).

Validity Accuracy and adequacy of the interpretations and explanations of researchers regarding their research object.

Valorisation Transferring and translating scientific knowledge (back) to practice, to make it of value for specific communities or our society in general.

Voluntarism Scientific position in which the free will of human beings is thought to fulfil a central function in social practices and in which individual actions are mainly viewed as the consequences of people's own intentions and choices.

References

Albertsen, N. and B. Diken, 2003. What is the social?. Lancaster University, Lancaster. Available at: http://comps.lanc.ac.uk/sociology/soc033bd.html. Accessed 13 November 2003.

Austin statement, 2003. Concept Statement Activist Anthropology at the University of Texas, Program in Activist anthropology. University of Texas at Austin version 23 October 2003, http://www.utexas.edu/cola/depts/anthropology/programs/activist/.

Beck, U., 1986. Risikogesellschaft. Auf dem weg in eine andere moderne, Frankfurt am main: Suhrkamp Verlag. Also in English: 1992. Risk society. Towards a new modernity (translation by Ritter, M.). Sage, London.

Benhabib, S. and D. Cornell, 1987. Feminism as critique: On the politics of gender. University of Minnesota Press, Minneapolis.

Beukema, L. and B. Valkenburg, 2007. Demand-driven elderly care in the Netherlands. A case of exemplarian action research. Action Research 5 (2): 115- 134.

Borda, O.F., 2006. The North-South convergence. A 30-year first-person assessment of PAR. In: Action Research 4 (3): 351-358.

Bryman, A., 2004. Social research methods. Oxford University Press, Oxford.

Butler, J., 1999. Gender trouble. Routledge, London, New York.

Callon, M., 1998. The laws of the markets. Blackwell Publishers, Oxford.

Cilliers, P., 1998. Complexity and postmodernism: understanding complex systems. Routledge, New York, London.

Coenen, H., 1987. Handelingsonderzoek als exemplarisch leren - een bijdrage aan de fundering van de methodologie van handelingsonderzoek. Konstapel, Groningen.

Coenen, H., 1996. The model of exemplarian action research. In: Boog, B., Coenen, H., Keune, L., and Lammerts, R. (eds.) Theory and practice of action research. Tilburg University Press, Tilburg, pp. 13-21.

Coenen H., 1998. On the foundations of a relationship based on equality between researcher and the researched party in exemplarian action research. In: Boog, B., Coenen, H., Keune, L. and Lammerts, R. (eds.) The complexity of relationships in action research. Tilburg University Press, Tilburg, pp. 17-36.

Côté, J. and C. Levine, 2002. Identity formation, agency and culture. A social psychological synthesis. Lawrence Erlbaum, Mahwah, New Jersey, London.

Denis, J-L. and J. Lomas, 2003. Convergent evolution: the academic and policy roots of collaborative research. Journal of Health Services Research and Policy 8 (Supplement 2): 1-6.

Derrida, J., 1983. The time of a thesis: Punctuations. In: Montefiori, A., Philosophy in France today. Cambridge University Press, Cambridge, pp. 34-51.

Durkheim, E., 1938. The rules of sociological method. Free Press, New York.

Foucault, M., 1982. Afterword - the subject and power. In: Dreyfus, H.L. and Rabinow, P., Beyond structuralism and hermeneutics. The Harvester Press, Brighton, pp. 208-227.

Gibbons, M., C. Limoges, H. Nowotny, S. Schwartzman, P. Scott and M. Trow, 1994. The new production of knowledge: the dynamics of science and research in contemporary societies. Sage, London.

Gibbons, M. and H. Nowotny, 2001. The potential of transdisciplinarity. In: J.T. Klein, W. Grossenbacher-Mansuy and R. Häberli (eds.), Transdisciplinarity: Joint problem solving among science, technology, and society. An effective way for managing complexity. Birkhäuser, Basel, pp. 67-80.

Giddens, A., 1976. New rules of sociological method. Hutchinson, London.

Giddens, A., 1979. Central problems in social theory - action, structure and contradiction in social analysis. MacMillan Press Ltd, London, Basingstoke.

Giddens, A., 1981. Agency, institution and time-space analysis. In: Knorr-Cetina K. and Cicourel, A.V., Advances in social theory and methodology. Toward an integration of micro-macrosociologies. Routledge & Kegan Paul, Boston, pp. 161-175.

Giddens, A., 1991. Modernity and self-identity. Polity Press, Cambridge.

Gieryn, T.F., 1995. Boundaries of science. In: Jasanoff, S. *et al.* (red.) Handbook of science and technology studies. Sage, Thousand Oaks, pp. 393-443.

Greenwood, D. and M. Levin, 1998. Introduction to action research. Social research for social change. Sage, Thousand Oaks.

Habermas, J., 1973. Wahrheitstheorien. In: H. Fahrenbach (ed.), Wirklichkeit und Reflexion, Festschrift für Walter Schulz, Pfullingen: Neske, pp. 211-265.

Habermas, J., 1984. Vorlesungen zu einer sprachtheoretischen grundlegung der soziologie. In: idem., Vorstudien und ergänzungen zur theorie des kommunikativen handelns. Suhrkamp Verlag, Frankfurt am Main, pp. 11-127.

Hagmann, J., E. Chuma, K. Murwira and M. Connolly, 1999. Putting process into practice operationalisation of participatory extension. AgRen Network Paper 94, ODI, London.

Haraway, D., 1989. Primate visions: Gender, race and nature in the world of modern science. Routledge, London, New York.

Harding, S., 1987. Is there a feminist method?. In: idem. (ed.) Feminism and methodology - social science issues. Indiana University Press/Open University Press Bloomington and Indianapolis/ Milton Keynes, pp. 1-14.

Heron, J., 1996. Co-operative inequity. Research into the human condition. Sage, London, Thousand Oaks, New Delhi, pp. 1.

Irigaray, L., 1985. This sex which is not one. Cornell University Press, New York.

Kristeva, J., 1991. Strangers to ourselves. Columbia University Press, Chichester, New York.

Kuhn, T.S., 1962. The structure of scientific revolutions. University of Chicago Press, Chicago.

Kuhn, T.S., 1970. Logic of discovery or psychology of research? and Reflections on my critics. In: Lakatos, I. and Musgrave, A., Criticism and the growth of knowledge. Cambridge University Press, Cambridge, pp. 1-24 and pp. 231-278.

Latour, B., 1993. We have never been modern. Harvard University Press, Cambridge.

Latour, B., 1998. To modernise or ecologise? That is the question. In: Braun, B. and Castree, N. (eds.) Remaking reality: Nature at the millennium. Routledge, London, New York, pp. 221-242.

Law, J., 2004. After method: Mess in social science research. Routledge, London, New York.

Levin, M and D. Greenwood, 2001. Pragmatic action research and the struggle to transform universities into learning communities. In: P. Reason and H. Bradbury (eds.), Handbook of action research. Participative inquiry & practice. Sage, London, Thousand Oaks, New Delhi, pp. 103-113.

Lyotard, J.F., 1992. The postmodern explained. Correspondence between 1982-1985. University of Minnesota, Minneapolis.

Midgley, G., 2000. Systemic thinking, philosophy, methodology, and practice. Kluwer Academics/ Plenum Publishers, New York.

Popper, K.R., 1963. Conjectures and refutations - the growth of scientific knowledge. Routledge and Kegan Paul Ltd, London.

Reason, P. and H. Bradbury (ed.), 2001. Handbook of action research. Participative inquiry & practice. Sage, London, Thousand Oaks, New Delhi, pp 1-15.

Rorty, R., 1985. Habermas and lyotard on postmodernity. In: Bernstein, R. (ed.) Habermas on modernity. Oxford University Press, Oxford, pp. 161-179.

Smaling, A., 2003. Methodologische objectiviteit en kwalitatief onderzoek (proefschrift), Lisse: Swets & Zeitlinger, 1987.

Slocum N., 2003. Participatory Methods Toolkit, A practitioner's manual, viWTA and King Baudouin Foundation Belgium.

Walliman, N, 2005. Your research project. Sage, London.

Webster's Ninth New Collegiate Dictionary, 1988. Merriam Webster Inc. Publishers, Springfield.

Weingart, P., 2000. Interdisciplinarity: the paradoxical discourse. In: Weingart, P. and Stehr, N. (eds.) Practising interdisciplinarity. University of Toronto Press, Toronto, pp. 25-41.

Contributing authors

Dominique Aerts obtained a Master's degree in engineering in agricultural and applied biological sciences in 1987 at the KULeuven (Belgium). For 11 years, she worked in and for Latin-America with a non-governmental organisation for development aid (ACT, now TRIAS). Subsequently she worked for 8 years at the Environment Administration in the Ministry of the Flemish Community, where she assumed several coordinating tasks involving the complete organisation and she led expert teams working in the field of policy making for integration themes, e.g. health and environment. For two years she has been manager of the science-policy interface at the Research Institute for Nature and Forest.
Affiliation: Research Institute for Nature and Forest (INBO), Belgium
E-mail: dominique.aerts@inbo.be

Conny Almekinders works in the chairgroup of Technology and Agrarian Development (TAD), the Social Science department of Wageningen University. She has a background in crop physiology and over the years her focus has shifted from agricultural crops to farmer, and farmer-researcher interaction. Seed systems and Participatory Plant Breeding (PPB) are still an important area of her interest. She combines this with exploring the field of research in action within academia. Over the last couple of years she coordinated the PhD programme on Participatory Approaches and Upscaling (PAU), a programme hosted by TAD.
Affiliation: Technology & Agrarian Development, Social Sciences Group, Wageningen University, the Netherlands
E-mail: conny.almekinders@wur.nl

Leni Beukema works at the department of Interdisciplinary Social Studies of Utrecht University since 1981. In 2005 she started her individual and independent research and training bureau: *LB: action learning and learning in action*. She worked on action research since the start of her academic career and has done research together with different trade unions in the Netherlands and a great variety of organisations. She teaches action research in many settings (students in social science at Utrecht University, PhD-students in the Netherlands and South Africa, managers of public organisations, workers in social welfare organisations, etc.).
Affiliation: Utrecht University (part time), independent researcher in LB (part time), the Netherlands
E-mail: lenibeukema@xs4all.nl

Maria de Bie holds a PhD in political and social sciences. She belongs since 1974 as a researcher to Ghent University. At the moment she is professor in educational sciences and social work. She is head of the department of social welfare studies. Her research topics are youth welfare, action research, community development, social care and welfare rights.

Affiliation: Vakgroep Sociale Agogiek - Department of Social Welfare Studies, Universiteit Ghent, Belgium
E-mail: Maria.Debie@ugent.be

Ludwine Casteleyn is researcher at the Center for Human Genetics at the University of Leuven and occupational health physician for Belgian federal civil servants. She studied medicine, occupational health, radioprotection and obtained a master's degree in Social Law. She was co-ordinator and participant in several European research activities on work and environment related diseases, ethics and privacy. She also was advisor to the Ministry of the Flemish Community, providing scientific support for policymaking in environment and health. Currently she leads activities on 'Ethics and communication' in the European Integrated Project NewGeneris and chairs the European expert group preparing the implementation of Action 3 of the European Environment and Health Action Plan, aiming at a more harmonised approach to Human Biomonitoring.
Affiliation: Center for Human Genetics at the University of Leuven, Belgium
E-mail: ludwine.casteleyn@med.kuleuven.be

Lieve Goorden has a background in political and social sciences (PhD University of Antwerp). She has been working at the Flemish Foundation for Technology Assessment (Stichting Technologie Vlaanderen) and at TNO in Delft (in the research group 'Strategy, Technology, Policy'). In October 1999 she joined the research group STEM at the University of Antwerp to coordinate the research cluster on technology assessment. In this research domain studies are undertaken on public debate and policy making on complex technological and social problems. She is professor and lecturer for the course on 'Technology and Society' in the Faculty of Applied Economics of the University of Antwerp. She is currently coordinator of the research project 'Nanotechnologies for tomorrow's society', with partners such as the University of Leuven, the Physics department of the University of Antwerp and IMEC (Interuniversity Micro-Electronic Centre).
Affiliation: University of Antwerp, Belgium
E-mail: lieve.goorden@ua.ac.be

Gemma van der Haar is assistant professor in Disaster Studies at Wageningen University. She is a development sociologist interested in the ethnography of political practice in conflict and post-conflict situations. She has extensive research experience on the Zapatista uprising in Chiapas, Mexico. Other research interests include local government and conflict, parallel governance and contested state formation, and conflicts over natural resources. She also works on methodological issues, especially the options and dilemmas in collaborative research.
Affiliation: Disaster Studies, Social Sciences Group, Wageningen University, the Netherlands
E-mail: gemma.vanderhaar@wur.nl

Jürgen Hagmann is a change management and organisational development facilitator. He has been working through action research and action learning on transforming rural extension and research/innovation services and NRM over the past 2 decades from local to international level. He has developed a range of methods and tools to operationalise action research/learning in rural development programmes in Africa, Asia and Latin America. Since 2005 he has been the team leader of PICOTEAM (www.picoteam.org), an organisation actively promoting action research approaches.
Affiliation: PICO, Southern Africa
E-mail: Jurgen.Hagmann@picoteam.org

Hans Keune is a Political Scientist working at the Faculty of Political and Social Sciences, University of Antwerp, Belgium. His expertise is mainly related to environment & health (science, social issues and policymaking), risk (perception, communication and assessment), knowledge (production, integration, interpretation and application), complexity, inter- and transdisciplinarity (boundary work science – government – society and participation), action research, group assessment/decision support methods (e.g. multi-criteria decision analysis, Delphi method, focus group).
Affiliation: University of Antwerp, Belgium
E-mail: hans.keune@ua.ac.be

Zacharia M. Mamabolo is a chief agricultural technician attached to Molemole Municipality in the Limpopo Department of Agriculture in South Africa. He was trained in participatory extension approaches (PEA) in 1998 and later became a trainer and mentor in this approach to subsequent generations of extension officers. He started the process of soil fertility management in selected villages from 1999 in his municipality and supported the dissemination of this innovation to other areas. He again served as a trainer of extension officers, scientist and farmers and provided mentoring support during the implementation of soil fertility management at community level.
Affiliation: Limpopo Department of Agriculture, South Africa
E-mail: deanm@idt.org.za

Michael H. Netshivhodza is an experienced facilitator who trained and mentored other officers in the implementation of PEA process and soil fertility management in five districts of Limpopo Province in South Africa. In 2004, he was seconded to provide a back-up support for the mobilisation of livestock farmers in 118 villages within Blougberg Municipality. From this process, he broadened the concept of local organisational development to its application a village. He is currently stationed at Madzivhandila College of Agriculture where he supports departmental projects by means of active facilitation as a registered facilitator, assessor and moderator under the ETDP SITA.
Affiliation: Limpopo Department of Agriculture, South Africa
E-mail: netshivhodzah@gmail.com

Jet Proost has a background in training and facilitation in innovation processes, where technical know how meets social change. After a career as assistant-professor at Wageningen University in Innovation Studies she now works as independent consultant for organisations in government and non-profit. In various contexts her strength lies in enabling people to discover their potential and overcome barriers to personal growth and collective action. Other roles have included active membership of various networks in sustainable development as well as being an IAF (International Association of Facilitators) Certified Professional Facilitator (CPF) and an assessor of the programme.
Affiliation: Jet Proost Communicatie Advies, the Netherlands
E-mail: jet.proost@planet.nl

Johannes M. Ramaru has for over a decade been championing processes supported by a GTZ funded programme that led to the development of participatory extension approaches (PEA) and innovation processes in Limpopo Province. He perceives himself as a facilitator for change at various levels with interest in innovation processes, local organisational development, stakeholder platforms for joint learning and development of competencies of extension staff, researchers and farmers for service delivery reforms. He is also a team member of PICO Southern Africa (www.picoteam.org).
Affiliation: Limpopo Department of Agriculture, South Africa
E-mail: mjramaru@yahoo.com

Rudi Roose educational sciences and criminology. He holds a PhD in educational sciences. He is a senior researcher at the department of Social Welfare Studies, Ghent University and associate professor at the department of Criminology, Free University Brussels. His research topics are child protection and youth care, children's rights and forensic social work.
Affiliation: Vakgroep Sociale Agogiek - Department of Social Welfare Studies, Universiteit Ghent, Belgium
E-mail: Rudi.Roose@ugent.be

Coyan Tromp is curriculum developer and teacher for the interdisciplinary bachelor track Future Planet Studies, a study in which solutions are sought for complex global issues such as climate change, food supply and water management. With a background in philosophy and psychology, in 2004 she finished her dissertation on the philosophy of social science. In the past, she has worked as consultant and researcher, experimenting with and teaching on subjects related to research in action. She also has been editor of the scientific journals of *Labour & Participation* and *Social Intervention*.
Affiliation: Institute for Interdisciplinary Studies, Faculty of Science, University of Amsterdam, the Netherlands
E-mail: j.c.tromp@uva.nl

Ben Valkenburg is lecturer in social studies at the University of Utrecht. He also works as a part-time consultant for Combo Consultancy, also in Utrecht. Through the years he has been involved in many research projects on industrial relations and labour-market policies. Most of these projects were based on the method of examplarian action research.
Affiliation: Faculteit der Sociale Wetenschappen, Universiteit Utrecht, the Netherlands
E-mail: B.Valkenburg@uu.nl

Gerard Verschoor's interests lie in the broader field of rural development in both Latin America and West Africa, with a special focus on agro-food studies and people-environment interactions. His research on agro-food studies focuses on the wider implications of 'the biotechnology revolution' for food production and consumption, and includes debates about GM food crops and biofuels, the revalorisation of local landraces and local knowledge, and the effects of standards and grades on relationships within the food chain. Research on people - environment interactions includes community-based conservation, local ecological knowledge, human adaptation to the environment, and the interaction between policy and the management of natural resources.
Affiliation: Rural Development Sociology, Social Sciences Group, Wageningen University, the Netherlands
E-mail: gerard.verschoor@wur.nl

Dirk Wildemeersch is a Public Health Officer working for the Flemish Health Authority. He graduated as a medical doctor at the Antwerp University and completed his Master's degree in environmental sciences at Ghent University. Since 1988, he has worked in Public Health; first in infectious diseases prevention as well as in environmental health. From 2002 until 2007 he coordinated the Environmental Public Health Section of the Flemish Health Authority and is now in charge of the Public Health Surveillance Division.
Affiliation: Flemish Agency for Care and Health, Public Health Surveillance Division, Belgium
E-mail: dirk.wildemeersch@wvg.vlaanderen.be

Keyword index

www.ingramcontent.com/pod-product-compliance
Lightning Source LLC
Chambersburg PA
CBHW061204240926
51846CB00006B/1110